T0233784

SpringerBriefs in Public Health

SpringerBriefs in Public Health present concise summaries of cutting-edge research and practical applications from across the entire field of public health, with contributions from medicine, bioethics, health economics, public policy, biostatistics, and sociology.

The focus of the series is to highlight current topics in public health of interest to a global audience, including health care policy; social determinants of health; health issues in developing countries; new research methods; chronic and infectious disease epidemics; and innovative health interventions.

Featuring compact volumes of 50 to 125 pages, the series covers a range of content from professional to academic. Possible volumes in the series may consist of timely reports of state-of-the art analytical techniques, reports from the field, snapshots of hot and/or emerging topics, elaborated theses, literature reviews, and in-depth case studies. Both solicited and unsolicited manuscripts are considered for publication in this series.

Briefs are published as part of Springer's eBook collection, with millions of users worldwide. In addition, Briefs are available for individual print and electronic purchase.

Briefs are characterized by fast, global electronic dissemination, standard publishing contracts, easy-to-use manuscript preparation and formatting guidelines, and expedited production schedules. We aim for publication 8–12 weeks after acceptance.

More information about this series at http://link.springer.com/series/10138

Germán Velásquez

Vaccines, Medicines and COVID-19

How Can WHO Be Given a Stronger Voice?

Germán Velásquez
Policy and Health
South Centre
Geneva, Switzerland

Originally published in France under the title:
VACCINS, MÉDICAMENTS ET BREVETS
La COVID-19 et l'impératif d'une organisation internationale
Copyright © L'Harmattan, 2021
www.harmattan.fr

ISSN 2192-3698   ISSN 2192-3701   (electronic)
ISBN 978-3-030-89124-4   ISBN 978-3-030-89125-1   (eBook)
https://doi.org/10.1007/978-3-030-89125-1

This Springer imprint is published by the registered company Springer Nature Switzerland AG
The registered company address is: Gewerbestrasse 11, 6330 Cham, Switzerland

Preface

The considerable health, economic and social challenge that the world faced in early 2020 with COVID-19 continued and worsened in many parts of the world in the second half of 2020 and into 2021.

Countries, international organizations, economic sectors, families and individuals have lived at a pace of adaptation, learning and innovation. The goal and challenge has been how to save people while preserving the economy and justice.

Many wanted to explore on their own, sometimes without listening to the major international health bodies, such as the World Health Organization (WHO), an agency of the United Nations system with long experience and vast global expertise and with members from every country on the planet. How can an agency like WHO be given a stronger voice to exercise authority and leadership?

This book is a collection of research papers produced by the author between 2020 and early 2021 that helps answer this question. The topics address the state of thinking and debate – particularly with regard to medicines and vaccines – that would enable a response to this pandemic or subsequent crises that may emerge. Specifically, the research papers ask the following questions:

- How to rethink pharmaceutical research and development (R&D) in the context of COVID-19?
- How can patents and access to vaccines and medicines be reconciled in the public interest?
- How can we rethink global and local manufacturing of medical products after COVID-19?
- How can we rethink the necessary reforms of the World Health Organization in the aftermath of COVID-19?

This book presents the South Centre's reflections and studies to provide policymakers, researchers and other stakeholders with information and analysis on issues related to public health and access to medicines and vaccines in the context of COVID-19.

Geneva, Switzerland   Germán Velásquez

Acknowledgements

The author is very grateful to Dr Carlos Correa for his useful suggestions and contributions. The author also greatly appreciates the inputs and contributions of Dr Paul Benkimoun, Dr Viviana Muñoz Tellez and Nirmalya Syam. Special thanks to Caroline Ngome Eneme for coordinating the publication process and editorial assistance.

Contents

About the Author

Germán Velásquez is, since 2010, the Special Adviser for Policy and Health of the South Centre in Geneva, Switzerland, which is an intergovernmental think tank of and for developing countries.

Until May 2010, he was Director of the World Health Organization (WHO) Secretariat on Public Health, Innovation and Intellectual Property, at the Director-General's Office, in Geneva. He is a pioneer in the new area on health, intellectual property and access to medicines, and he represented WHO at the World Trade Organization (WTO) Council for Trade-Related Aspects of Intellectual Property Rights (TRIPS) from 2001 to 2010.

Germán Velásquez graduated from Javeriana University, Bogotá, Colombia, with a degree in philosophy and humanities that was further complemented by a master's degree in economics and a PhD in health economics from Sorbonne University in Paris, France. In 2010, he received a PhD Honoris Causa in public health from the University of Caldas, Colombia. In October 2015, he received a Honoris Causa Doctorate from the Faculty of Medicine at Complutense University of Madrid, Spain.

Germán Velásquez is author or co-author of numerous publications on subjects such as health economics and medicines, health insurance schemes, globalization, international trade agreements, intellectual property, and access to medicines.

List of Abbreviations and Acronyms

ACT	Access to COVID-19 Tools (Accelerator)
AMC	Advance Market Commitment
API	Active Pharmaceutical Ingredient
BARDA	Biomedical Advanced Research and Development Authority (United States)
CEPI	Coalition for Epidemic Preparedness Innovations
CIPIH	Commission on Intellectual Property Rights, Innovation and Public Health
COVAX	COVID-19 Global Vaccine Access (Facility)
COVID-19	Coronavirus Disease 2019
DAP	Drugs Action Programme
EU	European Union
FCTC	Framework Convention on Tobacco Control
FENSA	Framework of Engagement with Non-State Actors
FPP	Finished Pharmaceutical Product
FTA	Free Trade Agreements
Gavi COVAX AMC	Gavi Advance Market Commitment for COVID-19 Vaccines
GPMB	Global Preparedness Monitoring Board
GSPOA	Global Strategy and Plan of Action on Public Health, Innovation and Intellectual Property (WHO)
HICs	High-Income Countries
IHR	International Health Regulations (2005)
IGWG	Intergovernmental Working Group
IOAC	Independent Oversight and Advisory Committee for the WHO Health Emergencies Programme
IP	Intellectual Property
IPPR	Independent Panel for Pandemic Preparedness and Response
LDCs	Least Developed Countries
LICs	Low-Income Countries
LMICs	Low- and Middle-Income Countries
LPP	Local Pharmaceutical Production

MICs	Middle-Income Countries
MSF	Médecins Sans Frontières
NGOs	Non-Governmental Organizations
PCT	Patent Cooperation Treaty
PHEIC	Public Health Emergency of International Concern
R&D	Research and Development
TAP	Tech Access Partnership
TRIPS	Trade-Related Aspects of Intellectual Property Rights
TWN	Third World Network
UHC	Universal Health Coverage
UN	United Nations
UNCTAD	United Nations Conference for Trade and Development
UNDP	United Nations Development Programme
UNHLP	United Nations Secretary-General's High-Level Panel on Access to Medicines
UNIDO	United Nations Industrial Development Organization
WHA	World Health Assembly
WHO	World Health Organization
WIPO	World Intellectual Property Organization
WTO	World Trade Organization

Chapter 1
COVID-19 Vaccines: Between Ethics, Health and Economics

1.1 Introduction

From 7 to 13 years of research and development (R&D) and 1.8 million clinical trials to develop a vaccine in the past, we have moved on to 10 months of R&D and tens of thousands of clinical trials to start vaccinating against COVID-19 in 2021.

One cannot talk about vaccines without referring to the Frenchman Louis Pasteur. In 1885, after 8 years of animal research, Pasteur announced the principle of vaccination: 'Inoculate weakened viruses which have the characteristic of never killing, giving a mild disease which preserves from fatal disease ...' (Institut Pasteur, 2021).

On 6 July 1885, a 9-year-old Alsatian boy, Joseph, bitten fourteen times by a rabid dog, gave Louis Pasteur the opportunity to test his treatment on humans. This first vaccination was a success, and the boy became the first human being to be vaccinated (Institut Pasteur, 2021). In 1908, in Lille, France, Albert Calmette began work on a vaccine against tuberculosis. Thirteen years later, the first baby was vaccinated in a Paris hospital. In 1948, the American Jonas Salk focused his research on a polio vaccine. Eight years later, after 1.83 million clinical trials, it was announced that Salk's vaccine was safe and effective in preventing polio (Hammond, 2020).

1.2 Development of the COVID-19 Vaccine

In response to the devastating COVID-19 crisis, the search for a vaccine led to an unprecedented and massive injection of public funds into global R&D. According to the WHO, there are more than 50 vaccine candidates in clinical trials around the world (WHO, 2021a). Laboratories in the US, Europe, Russia, China, Cuba and India have developed and are producing vaccines, several have been licensed and vaccination campaigns have been launched.

© SC: South Centre 2022

G. Velásquez, *Vaccines, Medicines and COVID-19*, SpringerBriefs in Public Health, https://doi.org/10.1007/978-3-030-89125-1_1

Basically, two types of vaccine technologies are used: the classical ones, based on the use of a whole, inactivated virus, or on the use of a part of the virus; and the so-called 'new' technologies, based on the use of 'pure' nucleic acid (DNA or RNA) (e.g. Moderna and Pfizer-BioNTech) or on the use of a viral vector (e.g., Oxford-AstraZeneca, Johnson & Johnson, Sputnik, CanSinoBIO) (Société de Pathologie infectieuse, 2021). These latter platforms have already been investigated for years in relation to other viruses. They have enabled the rapid identification of a vaccine against COVID-19 as soon as the infectious agent was identified and represent a revolutionary breakthrough.

Historically, large pharmaceutical companies have not been very interested in producing vaccines. Treatment of severe or chronic conditions is more cost-effective than prophylaxis (Bezat, 2021). However, the COVID-19 pandemic has changed this situation. The astronomical sums of public subsidies to private companies have changed their financial prospects and the global epidemic has created a huge potential market. It is a question of vaccinating the entire world's population; it is not known how many times and how often (Bezat, 2021).

1.3   Two Key Issues: Immunity and Contagion

Many questions remain open. The duration of immunity provided by the vaccine is one of them. If it provides immunity for 6 months, it would not be called a vaccine, but a drug. But it will take time to study the duration of vaccine immunity. Unfortunately, this time cannot be 'bought'. Simply injecting money into public health care, as Bill Gates may think, cannot solve everything, such as the necessary reconstruction of healthcare systems that have suffered years of budget cuts.

On the other hand, it is not yet known whether vaccines can block contamination of other people, which is fundamental to the concept of vaccination (Herzberg, 2021). Meanwhile, the emergence of variants complicates the situation, with a possible weaker immune response of some vaccines to certain variants.

At the same time, some vaccines will have a limited global reach due to their characteristics. For example, storing Pfizer's vaccine below −70 °C requires expensive refrigerators – more than 12,000 euros – that are not available in many countries, especially in remote areas, and complex logistics. US researcher William Haseltine also wonders whether 'Pfizer and Moderna have created a Lamborghini when what most countries really need is a Toyota' (Haseltine, 2020).

1.4   Vaccine Nationalism

By the end of March 2021, WHO reported more than 120 million cases and more than two million deaths worldwide . According to the WHO Director-General, as of 18 January 2021, 39 million doses of COVID-19 vaccine had been administered in

49 industrialised countries and only 25 doses in developing countries. 'Not 25 million; not 25 thousand; just 25' doses (WHO, 2021b). In the meantime, the situation has changed, but huge inequalities remain between the industrialised countries and the countries of the Global South.

Some governments, such as those of the United States of America, the United Kingdom and also the European Union, have wanted to buy (monopolise) the entire production of candidate vaccines, or prevent their export outside their borders, to cover their own population first and foremost, an operation known as 'vaccine nationalism'[1] (Santos Rutschman, 2020). The United States, for example, has signed at least six bilateral agreements, totalling more than one billion doses, more than enough to inoculate its entire population (328 million). The European Union (447 million), Britain (67 million) and Canada (37 million) have signed seven bilaterals each, with the potential to cover their populations two, four and six times over, respectively, according to Duke University's Center for Global Health Innovation (Serhan, 2020). Vaccine shortages, linked to production difficulties, have led not only to a fierce market, with uneven distribution, but also to geopolitical leverage games, a 'vaccine diplomacy'. For example, the Chinese vaccine Sinovac has reached Brazil, the Russian vaccine Sputnik has reached Argentina, and the Indian vaccine Covishield (with Oxford-AstraZeneca) has reached several countries in the Global South.

Vaccine nationalism is not new. In 2009, during the influenza A (H1N1) pandemic, similar 'nationalism' also emerged. Access to vaccines and treatments was determined by purchasing power, with high-income countries securing supplies for their populations ahead of the rest of the world.

1.5 The COVAX Mechanism

In June 2020, a global collaboration called the ACT Accelerator defined a financing mechanism for universal access to COVID-19 vaccines (called the COVAX mechanism). This global immunisation plan is co-led by Gavi (the Vaccine Alliance, an international organisation heavily influenced by the Gates Foundation), the Coalition for Epidemic Preparedness Innovation (launched in Davos in 2017) and the WHO (Gavi, 2020).

The announcement of COVAX generated a strong global response, especially from Southern countries concerned about equitable access to future vaccines. Almost a year later, the COVAX mechanism is being challenged because industrialised countries and big pharma have ignored the commitments made. Similarly, it has not been possible to open the debate on compulsory licensing. This World Trade

---

[1] This expression was used by the WHO Director-General Tedros Adhanom Ghebreyesus during a roundtable discussion on 6 August 2020.

Organization (WTO) legal mechanism would increase access to vaccines in countries but is strongly resisted by industrialised countries and industry.

Compulsory licensing, and even an exception as requested by South Africa and India and supported by many developing countries in 2020 and 2021 at the WTO, are certainly mechanisms to be used in times of a global pandemic threatening the entire planet.

1.6 Compulsory Licensing

The patent holder is free to exploit the invention protected by the patent or to allow someone else to exploit it. However, where justified by the public interest or the need to correct anti-competitive practices, the government may authorise a third party to use the invention, without the consent of the patent holder, under a compulsory license. The patent holder is thus obliged to tolerate the exploitation of his invention by a third party or by the government itself. In such cases, the public interest in ensuring wider access to the patented invention is considered more important than the patentee's private interest in fully exploiting his exclusive rights. Compulsory licenses therefore allow third parties to use an invention without the consent of the patent holder. For example, when certain drugs are protected by a patent and their price makes them unavailable to the local population, local pharmaceutical companies can obtain compulsory licenses to produce generic versions of patented drugs or to import generic versions of drugs from foreign manufacturers. Since 1995, there have been 108 compulsory licensing attempts for 40 pharmaceutical products in 27 countries (Velásquez, 2019).

1.7 Access to Medicines and Vaccines: A New Player

Historically, access to medicines has been in the hands of two actors: the commercial actor (pharmaceutical industry) and the health actor (ministries of health). COVID-19 introduced a new actor: the political actor (governments and opposition to governments). Today, governments buy and decide who should be vaccinated and when. However, they are at the mercy of industry, which makes the 'scientific' announcements about the efficacy of its products, announces timelines, sets expectations, imposes prices and demands immunity from possible negative side effects of its vaccines. Governments have less and less power to regulate and control the vaccine industry, or at least have so far demonstrated their inability to do so. The WHO watches helplessly and lucidly, as its recommendations are voluntary. If the primary objective of the WHO is public health, the industry seeks profit, the national health sector depends on the political actor, and the political actor seeks the votes (support) to stay in or gain power.

Industrialised countries may succeed in vaccinating all or three-quarters of their population by 2021, but they will probably ignore the ethical principles, health logic and economic rationality to which they committed to the COVAX mechanism. The purchase of individual vaccines has been left to the market of supply and demand. The concept of public goods advocated at the World Health Assembly in May 2020 by the UN Secretary-General and many heads of state and government seems to have been sidelined.

Many questions remain to be answered, such as the duration of vaccine coverage, its ability to block transmission and the medium- and long-term side effects. Or to what extent states will accept the industry's demand not to be responsible for possible side effects, to what extent contracts between industry and governments will be transparent, or whether common public goods will be patented. Every day more and more questions arise due to the speed at which the virus advances and solutions are proposed. COVID-19 clearly illustrates the need to use compulsory licensing and, ultimately, the question of how to implement a research and development (R&D) model for vaccines and medicines that ensures equitable access to health for all.

The management of a pandemic cannot be left to commercial companies competing with the primary intention of making money; the public interest needs to be placed well ahead of the commercial interest and knowledge needs to be in the public domain in the service of the progress of science.

References

Bezat, J. M. (2021). Covid-19: C'est par des biotechs que l'industrie pharmaceutique a produit un miracle: la mise au point en un an de plusieurs vaccins efficaces. *Le Monde*, 18 January 2021.

Gavi. (2020). The Bill & Melinda Gates Foundation. https://www.gavi.org/operating-model/gavis-partnership-model/bill-melinda-gates-foundation.

Hammond, A. (2020). Jonas Salk et le vaccin contre la polio – Les Héros du Progrès. *Contrepoints*, 23 February 2020. https://www.contrepoints.org/2020/02/23/364951-jonas-salk-et-le-vaccin-contre-la-polio-les-heros-du-progres-5.

Haseltine, W. A. (2020). COVID-19. Most countries needed a vaccine that could have been produced, stored and administered simply and at low cost. 23 December 2020. https://www.williamhaseltine.com/covid-19-most-countries-needed-a-vaccine-that-could-have-been-produced-stored-and-administered-simply-and-at-low-cost/.

Herzberg, N. (2021). Les vaccins anti-COVID évitent-ils la transmission du SARS-CoV-2? *Le Monde*, 6 January 2021.

Institut Pasteur. (2021). History: The final years 1877–1887. https://www.pasteur.fr/en/institut-pasteur/history/troisieme-epoque-1877-1887.

Santos Rutschman, A. (2020). How 'vaccine nationalism' could block vulnerable populations' access to COVID-19 vaccines. 17 June 2020. https://theconversation.com/how-vaccine-nationalism-could-block-vulnerable-populations-access-to-covid-19-vaccines-140689.

Serhan, Y. (2020). Vaccine nationalism is doomed to fail. *The Atlantic*, December 2020.

Société de Pathologie infectieuse. (2021). Vaccins contre la Covid-19: questions et réponses, Paris. 11 January 2021. https://www.infectiologie.com/UserFiles/File/groupe-prevention/covid-19/vaccins-covid-19-questions-et-reponses-spilf-24dec2020.pdf.

Velásquez, G. (2019). *Intellectual property and access to medicines: An introduction to key issues – Some basic terms and concepts.* Training Paper No. 1. South Centre, December 2019.

WHO. (2021a). COVID-19 vaccines. January 2021. https://www.who.int/emergencies/diseases/novel-coronavirus-2019/covid-19-vaccines.

WHO. (2021b). WHO Director-General's opening remarks at 148th session of the Executive Board. 18 January 2021. https://www.who.int/director-general/speeches/detail/who-director-general-s-opening-remarks-at-148th-session-of-the-executive-board.

Chapter 2
Medicines and Intellectual Property: 10 Years of the WHO Global Strategy

2.1 Introduction

This chapter examines the negotiating process and the steps given for the implementation of the World Health Organization (WHO) Global Strategy on Public Health, Innovation and Intellectual Property (resolution WHA61.21) 10 years after its May 2008 approval.

There was an impasse created by the developed countries' non-acceptance of the recommendations of the Report on Public Health Innovation and Intellectual Property produced by the Commission on Intellectual Property Rights, Innovation and Public Health (CIPIH) (2005), especially the recommendations related to Intellectual Property. To respond to that impasse, the 2008 World Health Assembly (WHA) created a group known as the Intergovernmental Working Group (IGWG). Following 2 years of protracted negotiations, almost unprecedented in the WHO, the IGWG developed the Global Strategy on Public Health, Innovation and Intellectual Property (GSPOA) that was adopted through resolution WHA61.21. Notwithstanding that the issues related to intellectual property were formulated in an ambiguous and inconclusive manner, this resolution undoubtedly was the broadest and most comprehensive mandate ever given to the WHO in the field of medicines.

As the elements of the GSPOA relating to intellectual property were unclear, the years following the adoption of the resolution were not easy to manage within the WHO. In 2011, the Director General established a Consultative Experts Working Group (CEWG) to study and recommend how to address those issues.

The CEWG considered that the system of research and development (R&D) primarily based on the grant of intellectual property rights was not delivering the

This chapter is largely taken from: Velásquez, G. (2019 December). *Medicines and Intellectual Property: 10 Years of the WHO Global Strategy*. South Centre Research Paper 100. https://www.southcentre.int/wp-content/uploads/2019/12/RP100_Medicines-and-Intellectual-Property-10-Years-of-the-WHO-Global-Strategy-_EN.pdf.

needed pharmaceutical products, particularly to address the diseases prevailing in developing countries. Thus, it recommended the beginning of negotiations for a binding instrument or treaty to finance R&D for pharmaceutical products. Today, this point seems far from the interest of the WHO Secretariat or the countries that promoted it simply because of the lack of space to continue discussing such a politically sensitive issue. If this is the case, the need for new ideas and models on how to prioritise, organise, and finance R&D for medicines remains valid.

The idea of a binding treaty embarrassed some developed countries that, in order to delay or replace it or as a condition to start a discussion on it (the reason was never totally clear), they introduced a long and disappointing exercise for undertaking R&D Demonstration Projects. At least 3 years were consumed (2012 to 2015) by this exercise. At the end of 2015 and having in view the delays in the implementation of the GSPOA, the United Nations Development Programme (UNDP) Secretariat in New York suggested that the Secretary-General of the United Nations create a High-Level Panel on Access to Medicines (UNHLP). On 14 September 2016 the Panel published a report that attempted to build on and advance the process of making medicines available and affordable within the spirit of the GSPOA.

The WHO Secretariat and developed countries had difficulty in accepting the report produced by the High-Level Panel. In the meantime, the GSPOA mandate ended in 2015. That same year, WHO Member States adopted resolution WHA68.18, which extended the GSPOA mandate from 2015 to 2022 and decided that an evaluation of the GSPOA implementation would be carried out in 2018 by an independent group of experts. This evaluation aimed at 'focusing on achievements, remaining challenges and recommendations on the way forward' (WHO, 2015).

The GSPOA Evaluation report and the overall review report were submitted to WHO Member States in November 2017. The GSPOA review report concluded, among other things, that the seven elements of the GSPOA remained valid although their implementation has been weak. The elements of the GSPOA were too general, and more focus was needed for the implementation of its recommendations.

To facilitate the implementation of the GSPOA and to align its execution with the WHO Thirteenth General Programme of Work (2019–2023), the Executive Board of the WHO decided on the elaboration of a Roadmap for the implementation of the GSPOA recommendations in the way suggested by the evaluation (2018).

The WHA 2019 requested Member States to take note of the Road Map. At the same time, several countries, including several developed countries led by Italy, proposed the resolution Improving the transparency of markets for medicines, vaccines and other health related technologies. The resolution, as adopted, begins by making reference to 'the Report by the Director-General on Access to medicines and vaccines (document A72/17) and its annex "Roadmap" for access to medicines, vaccines and other health products' (WHO, 2019). This issue was neither the most urgent nor the most logical and strategical for beginning the implementation of the Roadmap implementation. As is often the case under certain circumstances, pressure, or lobbies external to the WHO Secretariat, this means that things do not always proceed in their originally planned direction.

Chapter 2
Medicines and Intellectual Property: 10 Years of the WHO Global Strategy

2.1 Introduction

This chapter examines the negotiating process and the steps given for the implementation of the World Health Organization (WHO) Global Strategy on Public Health, Innovation and Intellectual Property (resolution WHA61.21) 10 years after its May 2008 approval.

There was an impasse created by the developed countries' non-acceptance of the recommendations of the Report on Public Health Innovation and Intellectual Property produced by the Commission on Intellectual Property Rights, Innovation and Public Health (CIPIH) (2005), especially the recommendations related to Intellectual Property. To respond to that impasse, the 2008 World Health Assembly (WHA) created a group known as the Intergovernmental Working Group (IGWG). Following 2 years of protracted negotiations, almost unprecedented in the WHO, the IGWG developed the Global Strategy on Public Health, Innovation and Intellectual Property (GSPOA) that was adopted through resolution WHA61.21. Notwithstanding that the issues related to intellectual property were formulated in an ambiguous and inconclusive manner, this resolution undoubtedly was the broadest and most comprehensive mandate ever given to the WHO in the field of medicines.

As the elements of the GSPOA relating to intellectual property were unclear, the years following the adoption of the resolution were not easy to manage within the WHO. In 2011, the Director General established a Consultative Experts Working Group (CEWG) to study and recommend how to address those issues.

The CEWG considered that the system of research and development (R&D) primarily based on the grant of intellectual property rights was not delivering the

This chapter is largely taken from: Velásquez, G. (2019 December). *Medicines and Intellectual Property: 10 Years of the WHO Global Strategy.* South Centre Research Paper 100. https://www.southcentre.int/wp-content/uploads/2019/12/RP100_Medicines-and-Intellectual-Property-10-Years-of-the-WHO-Global-Strategy-_EN.pdf.

G. Velásquez, *Vaccines, Medicines and COVID-19*, SpringerBriefs in Public Health, https://doi.org/10.1007/978-3-030-89125-1_2

Unlike the initial draft, first proposed by Italy, the approved resolution does not create any responsibility on the part of the WHO or Member States to ensure transparency of pharmaceutical R&D and clinical trials costs. In addition, the resolution urges Member States to take measures to disclose the price of medicines, an activity that the WHO has been performing for more than 10 years and for which a new mandate was not necessary. Rarely in the history of WHA drug-related resolutions have reactions been so different and contradictory. Some have claimed a great victory, others have called it a failure that, according to some, risked narrowing the mandate given by previous resolutions.

Another issue discussed in this chapter in the context of the GSPOA is the implementation of the 2014 resolution WHA67.21, Access to biotherapeutic products including similar biotherapeutic products and ensuring their quality, safety and efficacy. The resolution requested the WHO Expert Committee on Biological Standardization to update the 2009 guidelines, taking into account the technological advances for the characterisation of biotherapeutic products and considering national regulatory needs and capacities. This important issue is part of the Roadmap point VI, Regulatory system strengthening, where the deliverables are 'Guidelines, standards and biological reference materials to support decreased regulatory burden and support production and quality control of safe and effective health products' (WHO, 2019).

Finally, as noted, the GSPOA was renewed until 2022. It is unclear, however, what will happen in the next 2 years in the WHO in relation to medicines and the strategy follow-up. Will the implementation of the Roadmap follow an orderly and logical path according to global health priorities? Or will the WHO, each year, continue to deal with issues in the area of medicines that groups with different interests and objectives bring to discussion in the WHA, which has been the case in recent years?

2.2 The Background of the IGWG Negotiations

A third of the world's population does not have regular access to essential medicines, and this ratio reaches levels of half the population in certain developing countries. According to the 2019 Joint United Nations Programme on HIV and AIDS (UNAIDS) report, out of the 37.9 million people who should have received a retroviral treatment, only 23.3 million had access to the therapy at the end of 2018 (UNAIDS, 2019). At the end of 2018, 14.6 million people needed treatment compared with 6.8 million people in 2012 who did not receive treatment (UNAIDS, 2012). Medicines are one of the crucial tools for preventing, relieving or curing diseases. Having access to medicines is a fundamental component of the right to health as established by human rights treaties as well as by the constitutions in many countries (Seuba, 2008).

The financial burden of expenditures in medicines in most developing countries is borne by individuals and not by health insurers (private or public) as occurs in developed countries. In countries where the per capita income is less than $1000 per year, for instance, individuals as well as state governments cannot bear the cost of a second-line anti-retroviral treatment at $4000–$5000 per year. According to World Bank figures, one billion people currently live in extreme poverty (less than one dollar per day), and this is precisely the population which suffers the most serious health problems (World Vision, 2020).

Today it is recognised that the current patent protection system, as imposed by the Agreement on Trade-Related Aspects of Intellectual Property Rights (TRIPS), has a significant impact on the entire pharmaceutical sector and, more specifically, on medicine prices to an extent that may hamper access to medicines for the poorer populations in countries of the Global South. It is also alarming that the rules included in the TRIPS Agreement are not necessarily appropriate for those countries that are trying to meet health and development needs. Patents primarily determine new medicine prices. They grant exclusive protection for a minimum period of 20 years from the date of filing the patent application.

In its 2002 report, the United Kingdom Commission on Intellectual Property Rights (CIPR) recommended that countries 'ensure that their IP protection regimes do not run counter to their public health policies and that they are consistent with and supportive of such policies'. (Commission on Intellectual Property Rights, 2002). Although the TRIPS Agreement obliges World Trade Organization (WTO) members to provide patent protection for medicines, it also allows them to take certain public interest measures, such as compulsory licenses, parallel imports, exceptions to patent rights, rigorously defining patentability criteria, which may mitigate the impact of patent rights under certain conditions.

In 2006, the WHO Report on Public Health, Innovation and Intellectual Property Rights stated that 'the TRIPS Agreement allows countries a considerable degree of freedom in how they implement their patent laws, subject to meeting its minimum standards including the criteria for patentability laid down in TRIPS. Since the benefits and costs of patents are unevenly distributed across countries, according to their level of development and scientific and technological capacity, countries may devise their patent systems to seek the best balance, in their own circumstances, between benefits and costs. Thus, developing countries may determine in their own ways the definition of an invention, the criteria for judging patentability, the rights conferred on patent owners and what exceptions to patentability are permitted (…)' (WHO, 2006b).

During the May 2008 World Health Assembly, the WHO approved the Global strategy on public health, innovation and intellectual property (hereinafter GSPOA). The Global Strategy gave the WHO the mandate to 'provide (…), in collaboration with other competent international organizations technical support (…) to countries that intend to make use of the provisions contained in the Agreement on Trade-Related Aspects of Intellectual Property Rights, including the flexibilities recognized by the Doha Declaration on the TRIPS Agreement and Public Health (…)' (WHO, 2008, p. 43).

Developing countries that have tried to apply the flexibilities contained in the TRIPS Agreement, confirmed in different international fora, have been subjected to bilateral pressures (Smith et al., 2009, p. 687). The GSPOA recognised this problem and proposed technical assistance as one of the elements to overcome this obstacle: 'International intellectual property agreements contain flexibilities that could facilitate increased access to pharmaceutical products by developing countries. However, developing countries may face obstacles in the use of these flexibilities. These countries may benefit, inter alia, from technical assistance' (WHO, 2008, p. 34).

On the relationship between patents and the research and development of new medicines, one of the main arguments in favour of the use of patents in the pharmaceutical field is that they allow the financing of the research and development (R&D) of new products to address public health needs. However, a study carried out by the National Institutes of Health showed that, over a period of 12 years (1989–2000), only 15% of approved medicines were true innovations. According to Carlos Correa (2004), innovation in the pharmaceuticals field started declining just after the grant of patents for pharmaceutical products became generalised because of the TRIPS Agreement. He also pointed out that R&D on diseases which prevail in developing countries has been neglected. As Trouiller's well-known work pointed out, only 0.1% of all new chemical entities produced between 1975 and 1999 were for tropical diseases (Trouiller et al., 2002, p. 2188). Since then, a more recent analysis found that 'of the 850 new therapeutic products registered in 2000-2011, 37 (4%) were indicated for neglected diseases' (Pedrique et al., 2013). The so-called 'neglected diseases' seem to have been ignored rather than forgotten.

Tensions between public health and the new intellectual property rules introduced by the WTO TRIPS Agreement were epitomised by the lawsuits filed by 39 transnational pharmaceutical companies challenging South Africa's medicines law. The subject of access to medicines was submitted to debate by the WTO TRIPS Council in June 2001, and it concluded with the Doha Declaration on the TRIPS Agreement and Public Health (WTO, 2001). This Declaration was undoubtedly an important moment in this international discussion, but it did not provide a full-fledged solution. The inclusion of limitations on the use of the TRIPS flexibilities in the bilateral free-trade agreements (FTAs), which have been signed by several countries with the United States and later with the European Union, also increase the tension between public health and the international intellectual property rules.

It is in this tense international context that the WHA requested the WHO to set up the CIPIH to analyse the connections between intellectual property and access to medicines (WHO, 2003a, b).

As part of the 60 recommendations, the CIPIH report recommended that 'WHO should develop a global plan of action to secure more sustainable funding to develop new products and make products that mainly affect the developing countries more accessible' (WHO, 2006b, p. 187). Based on this recommendation, the 59th WHA approved resolution WHA59.24, which requested that an intergovernmental working group open to all WHO members be established.

The resolution requested the intergovernmental working group to report on the progress made to the 60th WHA through the Executive Board. The resolution also requested that the Director-General include in the intergovernmental group organisations of the United Nations (WHO, 2006b, para. 3.2., and 4.2.) non-governmental organisations (NGOs) in official relations with the WHO, expert observers, and public and private entities.

The intergovernmental group held negotiations for almost 2 years, between December 2006 and May 2008, with three meetings in Geneva, which were attended by over 100 countries, and several other meetings in all the WHO regions. This document intends to provide a view and describe the mistakes made as well as the failures of the process so that those who tell the story, as seen through rose-coloured glasses, are not the only ones to narrate the events.

2.3   The IGWG Stakeholders

The WHO Member States were obviously the main stakeholders in the negotiations of the Global Strategy. As it usually happens in United Nations negotiations, there were groups, alliances, and mediators which helped to build consensus.

A first group, led by the United States and Switzerland, was supported by Australia, Japan, South Korea, Colombia, and Mexico, and in some way, Canada. A second group, which was led by Brazil, Thailand, and India, was supported by a great majority of the developing countries, including discreet support from China. The European Union, which spoke with one voice, was led by Portugal during the first part of the IGWG and then by Estonia in their capacities as presidents of the European Union. Although the European Union did at certain times try to act as an intermediary between the countries of the first and second group, this role was eventually taken up by the Norwegian delegation, which actively worked to build consensus.

As far as the role played by countries is concerned, the cohesion of the African Group should be pointed out since it spoke with one voice in coordination with the rest of the developing countries in most cases, such as during the WTO Doha Ministerial Conference discussion in 2001 on TRIPS and access to medicines.

The NGOs and non-profit organisations in the public health field played an important role. The role the NGOs have played in promoting access to medicines in the WHO governing bodies is well known and recognised (Velásquez, 2011). Maybe because of the enthusiasm generated by the negotiations, some organisations abandoned their discreet and effective lobbying for an open and visible promotion of certain issues, which did not always help the public health agenda to move forward or to build consensus.

The pharmaceutical industry, perhaps fearing the negotiations' scope and sensing the risk of having its commercial interests affected in the long-term – in particular with regard to intellectual property – was permanently present in the hallways and corridors, actively and ostentatiously trying to influence the different stakeholders. More than 80 industry representatives (associations and private industries) were present at the Palais des Nations in Geneva during the 2008 World Health Assembly.

Academia: An initiative such as that of the IGWG, which led to the adoption of the Global Strategy, was closely followed and analysed by academia. University professors from different parts of the world gave their opinion and tried to develop the issues addressed by the IGWG, no doubt bringing vision and analysis with greater depth than the flow of discussions within the United Nations.

Other United Nations agencies: Unfortunately, several United Nations agencies that fully share a public health vision, such as the United Nations Children's Fund (UNICEF), UNDP, and UNAIDS, were practically absent from the discussion. The WIPO and WTO participated throughout the negotiations, and the group of industrialised countries as well as the Secretariat of the WHO requested their comments and points of view on subjects related to the interpretation and management of intellectual property.

The WHO Secretariat was at first disoriented – a situation which led to the failure of the first IGWG meeting. The Director-General and the Deputy Director-General particularly invested their efforts fully in monitoring and supporting the negotiating process. According to some Geneva observers of the IGWG process, the Assistant Director-General who covered this topic had to leave the Organization mostly due to the failure of the first meeting, and a special PHI group (Secretariat of the WHO for Public Health, Innovation and Intellectual Property) was created in the Office of the Director-General. Many technical departments of the WHO, such as the Special Programme for Research and Training in Tropical Diseases (TDR) or the Department of Ethics, Trade and Human Rights, closely followed the discussions. The Department of Essential Medicines, which was the birthplace of the discussion, kept some distance. The WHO regional consultants in the field of medicines followed the negotiations as if it were their own.

2.4   The IGWG Process

2.4.1   The First Meeting in Geneva: 4–8 December 2006

The online consultation that took place before the meeting regarding the draft prepared by the Secretariat gave an indication of the controversial topics which would appear throughout the negotiations. Thirty-one contributions from different countries, industries, academia, and NGOs were received. The subject of a possible international convention or treaty on research and development of new products as an alternative system to that of the patented medicines, as the primary or even sole source of R&D funding, was undoubtedly the main subject of disagreement between the negotiating parties. The issue of whether to include the concept of access to treatment as a human right also made certain delegations nervous.

The six elements of the strategy to be presented by the WHO Secretariat at the first meeting were: (1) priorities of the requirements in terms of R&D, (2) identification of the flaws in the research agenda, (3) promotion of R&D, (4) building and improving the capacity for innovation, (5) improving access, and (6) ensuring sustainable funding mechanisms. The issue of intellectual property, which should have been a common denominator between these six elements, had practically disappeared. During the chaotic discussions, which characterised the entire meeting, the group of developing countries managed to reach general acceptance of the need to reintroduce the issue of intellectual property. The WHO Secretariat, probably due to pressure from certain Member States, decided to isolate this issue in a separate chapter (now element 5: Application and management of intellectual property to contribute to innovation and promote public health.). This constitutes the first and perhaps the most fundamental problem of the negotiations. Due to insistence primarily from the African Group, a second element regarding transfer of technology was included (element 4 of the approved strategy).

Speaking of the African Group, the organisation and coherence of their well-prepared interventions was the most positive aspect of this first meeting. Another point which the developing countries achieved was to include the possible negative impact of the free-trade agreements along with their requirements that go beyond the TRIPS requirements, known as the TRIPS-plus measures.

It was clear during the discussions that for most of the developing countries the new intellectual property rules required by TRIPS and the free-trade agreements are a negative factor with regard to access to medicines and innovation in the developing world. On the other hand, a small group of industrialised countries defended the position that the problem does not lie in intellectual property rights and patents but rather in the lack of funding, defective health infrastructures, and lack of political will. During the meeting (and practically throughout the negotiations), this same group of countries questioned the authority of the WHO in the area of intellectual property, insisting that this was an issue that should be dealt with by the WIPO and the WTO. According to these countries, the WHO should only be involved in health-care aspects (WHO, 2007b, paras. 20, 21 and 31). excluding other decisive aspects influencing the health sector. Agreement could not be reached on the inclusion of a reference to human rights as well as whether to state that public health has priority over intellectual property rights.

2.4.2 Regional Consultations

Regional and inter-country meetings took place during the second semester of 2007 throughout the WHO regions – AFRO (Regional Office for Africa) in the Congo; AMRO/PAHO (Regional Office for the Americas/Pan American Health Organization) in Washington, DC, USA; Bolivia; Rio de Janeiro, Brazil; and Canada; EMRO (Regional Office for the Eastern Mediterranean) in Egypt; EURO (Regional Office for Europe) in Serbia; SEARO (Regional Office for South East Asia) in the Maldives; and WPRO (Regional Office for the Western Pacific) in the Philippines.

Undoubtedly, the most relevant meeting was the one in Rio de Janeiro, which produced what was referred to as the Rio document and had the greatest influence on the final strategy document. The countries that took part in the meeting were Argentina, Brazil, Chile, Costa Rica, Cuba, Ecuador, El Salvador, Honduras, Mexico, Peru, Suriname, Uruguay, and Venezuela. The originality and correct choice of the Rio document was to try to include a context, a goal, and a set of principles based on human rights in the strategy. The Rio document's 11 principles gave a vision and, in a way, unveiled the philosophy of how the problem should be approached. The first three principles showed the spirit behind this document:

1. The right to health protection is a universal and unalienable right, and it is the governments' obligation to guarantee that the instruments to implement it are available.
2. The right to health takes precedence over commercial interests.
3. The right to health implies access to medicines.

Although the only regional consultation officially organised by AMRO/PAHO was the one in Ottawa, Canada, on 22–23 October 2007, this consultation limited itself to debating some controversial points contained in the Rio document. Canada was especially opposed to including items from the Rio document, particularly the reference to human rights. Another point that was contested by the North American countries was the WHO leadership in actions related to intellectual property. They also attempted to restrict the strategy's scope to three diseases – malaria, tuberculosis and AIDS – like in the old Doha discussions. Some of the participants at the meeting in Canada insisted on the United Nations (UN) technique, which consists of solving controversies by looking for a previously agreed-to text.

Between 15 August and 30 September 2007, the WHO Secretariat organised the second round of contributions through its webpage. Sixty-five contributions were received from governments, national institutions, NGOs, academics, patient associations, and the pharmaceutical industry (WHO, 2007a, para. 11). 'The unmanaged nature of Web-based hearings' was a problem for many.[1] Indeed, in the second public consultation, the number of presentations supporting a strong intellectual property protection increased enormously. This was questioned by many NGOs, which pointed out that the industry was distorting the spirit and the aim of the IGWG (Wibulpolprasert et al., 2007, p. 1754).

This second round was characterised by the richness of the proposals. The focus was on the very intense discussion on intellectual property and the possible alternative mechanisms for funding R&D for pharmaceutical products, resulting in the formation of two groups. The first group promoted proposals, such as the treaty on R&D, incentives, 'patent pools', or 'advance market commitments'.[2] The second

---

[1] Forman. L. Desk review of the intergovernmental working group on public health, innovation and intellectual property from a right to development perspective. Unpublished paper, Geneva, March 2009.

[2] Frederick M. Abbott and Jerome H. Reichman, 'Strategies for the Protection and Promotion of Public Health Arising out of the WTO TRIPS Agreement Amendment Process', Florida State

group, which was led by the industry and certain United States institutions, preferred solutions based on the market, arguing that a strong intellectual property
protection is the best incentive for stimulating R&D.[3] Some proposals, such as that
of the Italian alliance for the defence of intellectual property, challenged the role of
the WHO in this field, arguing that this role belonged exclusively to the WTO
and WIPO.[4]

2.4.3   Second Meeting, 5–10 November 2007

As a result of the regional and inter-country exercises, interest in the discussions
increased to the point that the number of countries represented at the second meeting of the IGWG reached 140, with 18 NGOs, 11 experts, and four or five United
Nations specialised agencies. Two working groups were created on elements 5 and
6 of the strategy (management of intellectual property and improving access), as
well as a subgroup, which started working on the plan of action.

Surprisingly enough, point 30.2.3.c – 'encourage further exploratory discussions
on the utility of possible instruments or mechanisms or essential health and biomedical research and development, including, inter alia, an essential health and biomedical *research and development treaty*'[5] – was approved at this second meeting
(WHO, 2008, p. 39). Undoubtedly, this was one of the central and most important
points of the Global Strategy that the industry as well as some industrialised countries were most opposed to. It is possible that the Chinese delegation support at this
point was the deciding element for the idea of a possible international treaty for the
funding of pharmaceutical R&D to be agreed upon at the end of the meeting, leaving only the determination of the role of WHO pending, which remained in
parentheses in the stakeholders' column. One-and-a-half years later, at the January
2009 Executive Board and the 2009 WHA, a group of nine countries, with the presence of the WHO Secretariat acting as an observer, used the WTO green room tech-

---

University and Duke University; James Love, Knowledge Ecology International; Itaru Nitta,
Green Intellectual Property Scheme System to impose a levy on patent applicants to establish a
trust fund to facilitate eco-Aidan Hollis, A Comprehensive Advanced Market Commitment;
Thomas Pogge, Track 2.

[3] Jeremiah Norris, Hudson Institute, USA; Harvey Bale, IFPMA; Ronald Cass, Centre for the Rule
of Law; Wayne Taylor, Health Leadership Institute, McMaster University; Anne Sullivan,
International Association for Business and Health; Hispanic-American Allergy Asthma and
Immunology Association; the National Grange of the Order of Patrons of Husbandry; International
Chamber of Commerce; Healthcare Evolves with Alliance and Leadership; and US Chamber of
Commerce.

[4] Daniele Capezzone, Benedetto Della Vedova, Veaceslav Untila and Kelsey Zahourek, Government
Institution, European Parliamentarians and the Property Rights Alliance, Italy; Harold Zimmer,
German Association of Research-based pharmaceutical manufacturers; and Ronald Cass, Centre
for the Rule of Law.

[5] Emphasis added.

nique and agreed to exclude the WHO as one of the stakeholders of this activity of the plan of action. This was perhaps the most serious flaw of the entire negotiations since it showed not only a refusal to study truly innovative solutions to a fundamental problem, but it also seemed to indicate that there was no clear vision on the future of access to medicines.

2.4.4 Continuation of the Second Meeting of the IGWG: 28 April to 3 May 2008

'This is the same meeting, let's go on as if this had just been a weekend recess' repeated the WHO Secretariat over and over again, but the weekend had lasted 6 months. Negotiations resumed with 147 registered Member States, 11 experts, over 20 NGOs, and United Nations specialised agencies. After negotiating one sentence at a time, and sometimes even one word at a time, consensus was reached on four of the seven elements. The remaining elements were element 4: transfer of technology, element 5: management of intellectual property, and element 6: improving delivery and access.

Many of the open points in parentheses pending consensus had been blocked only by the United States, and several countries requested that 'pending USA approval' be indicated on the draft with respect to these elements. The most problematic element for the United States delegation was element 5 in aspects such as the need to find new incentive schemes for research, the role of the WHO with regard to intellectual property, protection of test data, and the reference to TRIPS-plus provisions in bilateral trade agreements.

2.4.5 Sixty-First World Health Assembly, 24 May 2008

During the 61st World Health Assembly, practically a third meeting of the IGWG was held. In fact, it was somewhat like a parallel World Health Assembly since most of the countries participating in the assembly also took part in the negotiations, to the extent that some countries with small delegations preferred to be present at the IGWG negotiations and not at the 'normal' Assembly activities. During the week of the WHA, the eight working hours of the day were not enough, and beginning Wednesday, 28 May 2008, night sessions took place. In the last day, the activities went on until 3:00 a.m.

For the first time in 2 years of negotiations, on the Friday before the close of the Assembly, the WHO Secretariat authorised a WTO green room-type meeting (a closed-door meeting with a group of nine countries). This was initially called by the President as a lunch with 'the President's friends', which then went on as a simple closed-door meeting until 5:00 p.m. This practice, the first one in the history of the

WHO (except for some negotiations on the anti-tobacco convention) was strongly criticised by many countries in public, and they even threatened not to recognise the consensus reached by the nine countries in the green room at the 2008 WHA plenary session. The criticism from the delegations was even stronger during the 62nd WHA in May 2009 when they found out that another round of green room negotiations took place to solve the problems with the issues in parentheses that were pending. This round of negotiations led, as already mentioned, to the exclusion of the WHO as a stakeholder in the activity related to a treaty on R&D.

Several developing countries (Argentina, Bangladesh, Barbados, Bolivia, Cuba, Ecuador, Ghana, India, Jamaica, Nicaragua, Suriname, and Venezuela) expressed their disagreement with the way the closed-door informal consultations were carried out as well as with the result of these consultations to exclude the WHO as a stakeholder in future discussions regarding a possible international treaty.

On the last day and last moment of the Assembly (2009), a resolution sponsored by Canada, Chile, Iran, Japan, Libya, Norway, and Switzerland was approved with the support of the United States. This resolution referred to an approved document A62/16 Add.3, which excluded the WHO from future discussions regarding the treaty. It is important to point out that many of the main stakeholders during the 2-year negotiations, such as Brazil, India, Thailand, Philippines, or the African Group, did not cosponsor this resolution. It is also somewhat surprising that countries, such as Japan, who were absent from the negotiations or whose participation was rather low profile during the negotiations, appeared at the last moment as cosponsors of the resolution.

In an open letter to the WHO Member States, dated 18 May 2009, seven NGOs (Essential Action, Health Action International, Health Gap, Knowledge Ecology International, Médecins Sans Frontières, Oxfam International, and Third World Network) stated that we wish to 'call your attention on the document A62/16 Add.3 where the results of informal consultations among some Member States on stakeholders are presented. We are surprised that WHO has been removed as stakeholder in action 2.3(c) that "encourage further exploratory discussions in the utility of possible instruments or mechanisms for essential health and biomedical R&D, including inter alia, an essential health and biomedical R&D treaty." (…)WHO is the UN agency with the global mandate for health. It is unacceptable that there would be any opposition to the WHO having a role in this discussion' (KEI, 2009). Further on, the seven NGOs indicated that such a decision would go against the spirit of resolution WHA61.21 (see WHO, 2008, Agenda item 11.6).

2.5   The Global Strategy and Plan of Action on Public Health, Innovation and Intellectual Property

From 1996 to 2008, 12 WHA resolutions have referred to intellectual property and access to medicines. The mandate given to the Assembly can be summarised in two points:

1. Monitor the impact on health of the international trade agreements.
2. Support countries in formulating policies and measures intended to optimise the positive aspects and to lessen the negative impact of these agreements.

The GSPOA (WHO, 2008), which was approved by the WHA in May 2008, confirmed and extended the previous mandate given by the 12 WHA resolutions on the WHO involvement in public health and intellectual property.

2.5.1   Main Elements of the 2008 Global Strategy

(a) The strategy recognises that the current initiatives to increase access to pharmaceutical products are insufficient.
(b) It also recognises that the incentive mechanisms of the intellectual property rights are not delivering for people living in 'small or uncertain potential paying markets'.
(c) The GSPOA recognises that the current system of innovation based on the incentive provided by intellectual property has failed to stimulate the development of drugs for diseases that disproportionately affect the majority of the world's population living in developing countries.
(d) While it does recognise the role of intellectual property, the Global Strategy specifically recognises that 'the price of medicines is one of the factors that can impede access to treatment'.
(e) There is no restriction on the scope in terms of diseases or products as was negotiated in Doha and in the IGWG process.
(f) It recognises that the 'international intellectual property agreements contain flexibilities that could facilitate increased access to pharmaceutical products by developing countries. However, developing countries may face obstacles in the use of these flexibilities'.
(g) The Global Strategy aims to promote new thinking on innovation and access to medicines.
(h) The strategy also recognises that the public policies to promote competition can contribute to the reduction of the price of medicines.
(i) Paragraph 2.3(c) of the GSPOA refers to 'encourage exploratory discussions' on a possible international treaty on research and development of new pharmaceutical products.

2.5.2   Additional Mandates of the 2008 Global Strategy

(a) To 'strengthen education and training in the application and management of intellectual property, from a public-health perspective (…)' (WHO, 2008, point 34).
(b) To establish urgently an expert working group (EWG) to examine proposals for new and innovative sources of funding for research and development of pharmaceutical products (WHO, 2008, para 7). These 'new and innovative sources' included a possible binding treaty on how to finance R&D for pharmaceutical products.

2.5.3   Progress in the Implementation of the GSPOA

The progress[6] made in implementing the Global Strategy and its action plan is thus far limited to four points:

1. Patent Pools.[7] These were one of the many elements of the mandate given to the WHO by resolution WHA61.21. Patent pools can facilitate equitable access and make new HIV treatments more affordable. They can also facilitate the development of new fixed-dose combinations suitable to address developing countries' treatment needs. Patent pools may consist of compulsory licenses or licenses voluntarily granted by the patent holder as is the case of the current Medicines Patent Pool (MPP) created with funds from the French initiative, UNITAID. Voluntary patent pools do not constitute a structural solution to the access to medicines problem.
2. The so-called Platform on Innovation, promoted by the Pan American Health Organization (PAHO). It is a kind of 'Facebook of medicines', a virtual network reporting on various activities in the pharmaceutical field.
3. Collaborative activities between the WHO, WTO, and WIPO which led to the so-called 'tripartite report', Promoting Access to Medical Technologies and Innovation (WTO, WIPO, WHO, 2012). Whereas the study could represent progress for the WTO and WIPO, given that it talks about the TRIPS flexibilities with no taboos, it does not reflect the fact that the WHO was the international organisation that had, until then, led this issue. There are 20 WHA resolutions referring to intellectual property and public health, adopted between 1996 and 2012, and most of these resolutions are cited by the report in a table on page 44. These resolutions clearly have a prescriptive character for the WHO Secretariat on how to preserve public health from the potential negative impact of new

---

[6] A Canadian private firm contracted by the WHO conducted an evaluation of the global strategy. The results say very little, since the terms of reference were poorly drafted.
[7] http://www.medicinespatentpool.org/.

international trade rules. Numerous WHO publications on this topic published over the past 15 years also point in this direction.

The disclaimer of the report states that '(...) the published material is being distributed without warranty of any kind, either expressed or implied. The responsibility for the interpretation and use of the material lies with the reader. In no event shall the WHO, WIPO and WTO be liable for any consequences whatsoever arising from its use'. This type of disclaimer may give the reader the misleading impression that the WHO has no opinion as to whether, for instance, a compulsory license may promote access to medicines in particular circumstances, or whether an international exhaustion regime that allows parallel imports from any country can reduce medicines costs and, therefore, contribute to access. The 20 resolutions mandate the WHO to engage, promote, and defend mechanisms and policies in favour of access.

The trilateral report is unambitious and does not reflect the work that the WHO has carried out under its mandate. It is curious that this 251-page document has no single recommendation – not even a conclusion.

4. Demonstration projects, an idea launched and promoted by the European Union at the WHO. These demonstration projects, which were not part of the existing mandate in the GSPOA or in the various resolutions of the World Health Assembly, were used to postpone the start of negotiations on a binding R&D treaty. In 2012 and 2013, project selection took place in a process that involved the six WHO Regional Offices. This selection process was heavily criticised by non-governmental organisations and some observers as a distraction to delay the start of negotiations on a binding treaty.

On 30 September 2014, a meeting convened by France, Switzerland, South Africa, and the WHO Secretariat was held at the Palais des Nations in Geneva to discuss and announce how and by whom the demonstration projects would be funded. It was attended by 15 developed and six developing countries. The WHO Secretariat presented the financial situation for the implementation of the projects: the estimated cost for 4 years was $50 million, of which $3 million had been received ($2 million was a donation from France, which was given directly to Drugs for Neglected Diseases initiative (DNDi) and not to the WHO Secretariat). The meeting ended in an impasse as developed countries stated that they would only announce their funding pledges after 'non-traditional donors' announced theirs. This concept of non-traditional donors has recently been introduced by developed countries to promote the idea of emerging countries participating as donors. South Africa simply announced that the BRICS countries (Brazil, Russia, India, China, and South Africa) would consider funding. The African countries present expressed concern about the Ebola epidemic and insisted that this was the priority for them in terms of new financial contributions.

More than 6 years after the approval of the demonstration projects, the required funding was not received. The start of negotiations for a Convention was not formally contingent on the results of the demonstration projects but, in practice, the debate on such projects took so much space that the start of negotiations was set aside. If the demonstration projects were only a pretext for delaying the subject of a treaty, as many suspected, they were certainly successful as the proposed treaty was not only delayed but virtually removed from the WHO agenda.

2.5.4   The Collaboration of the WHO with Other International Organisations

Interestingly, the United Nations agencies invited to participate in the debates on intellectual property and health, which took place in the WHO between 2008 and 2018 in the context of the GSPOA, were the WIPO and WTO. This is despite the fact that there are other United Nations agencies that are much closer to the work of the WHO, such as the UNDP, UNAIDS, the United Nations Conference on Trade and Development (UNCTAD), and the Commission on Human Rights. These were not invited by the WHO to participate in the discussions on the subject of access to medicines. In the case of the UNDP, its presence at the country level has been much more relevant in recent years than the rest of the agencies mentioned above.

The dialogues or cooperation between the WHO, WIPO and WTO from 2010 to 2015 have placed the international debate on access to medicines in limbo. This was undoubtedly one of the reasons why the UNDP sought to rescue the issue by suggesting that the United Nations Secretary-General convene a High-Level Panel on access to medicines by the end of 2015. The High-Level Panel released its report on 14 September 2016 as discussed later in this chapter.

2.6   The WHO Consultative Expert Working Group

As the IGWG faced opposition from industrialised countries on the idea of an international convention or treaty on biomedical R&D, the 2008 WHA created a group of experts – the Expert Working Group (EWG) – to analyse and recommend what to do on this issue. The report of the EWG mandated by resolution WHA61.21 failed to address the issue of intellectual property and was rejected by the WHO Member States. The report of this group was strongly criticised at the WHO Executive Board in January 2010, following a complaint by Dr. Cecilia Lopez, one of the members of the group. After the WHA's rejection of the EWG report in 2010, another EWG was requested by a WHA resolution the same year. At the beginning of 2011, the WHO Director-General established a WHO Consultative Expert Working Group (CEWG) to address the financing of R&D in the context of intellectual property issues. In July 2011, the Chair of the CEWG announced that 'the CEWG will recommend to the 2012 World Assembly the commencement of formal intergovernmental negotiations for the adoption of a binding international treaty on R&D for health' (Rottingen, 2011).

2.6.1 A Binding International Convention

The CEWG report contained several findings and recommendations. The finding that the current system of incentives through the protection of patents has failed to respond to the problems of the developing countries, where most of the world population lives was a clear starting point.

In fact, on sustainable long-term access to medicines for developing countries and today even for developed countries, it is clear that rather than recommend, the WHO should use its capacity to legislate. A convention or an R&D treaty is undoubtedly the path to follow. Under Article 19 of the WHO Constitution '(T)he Health Assembly shall have authority to adopt conventions or agreements with respect to any matter within the competence of the Organization. A two-thirds vote of the Health Assembly shall be required for the adoption of such conventions or agreements, which shall come into force for each Member when accepted by it in accordance with its constitutional processes' (WHO, 2006a). Despite the notorious regulatory powers its constitution confers, 'WHO has paid but scarce attention to law – especially the *hard law* – as a tool to protect and promote health. On the contrary, the Organization has shown itself to be more in favour of seeking a political agreement and has excused itself in its medico-sanitary profile in order to take on more of a health care than a legal role' (Seuba, 2008). Article 19 of the WHO Constitution was only used once in the 70 years of the Organization's existence: the Framework Convention on Tobacco Control (FCTC).

In May 2012, the WHO Member States met at the World Health Assembly in Geneva. They adopted resolution WHA 65.22 endorsing the recommendations of the CEWG, which for many of the WHA participants and observers meant the first step towards a change in the current pharmaceutical research system. Arguing that the market is not enough to drive R&D, the CEWG recommended the negotiation of an international convention in which all countries would commit to promoting R&D: 'formal intergovernmental negotiation should begin for a binding global instrument for R&D and innovation for health' (World Health Assembly, 65., 2012).

The aim of an international convention would primarily be to set up an international public fund for pharmaceutical R&D. To ensure sustainability of the fund, the convention would need to provide for a compulsory contribution by signatory countries according to their level of economic development. In return, the products and results financed by this fund would be considered as public goods benefiting all these countries. Hence, the idea is not a new financial contribution but rather an innovation model which focuses more on patients' interests than does the current system. Moreover, the costs of research activities financed by this public fund would have to be transparent to guarantee a more efficient and less costly medical innovation system that meets the real sanitary needs of countries of both the Global North and the Global South.

A binding international convention, negotiated under the auspices of the WHO, could thus serve to sustainably finance R&D on useful and safe medicines responding to the needs of all patients and available at prices accessible to patients and health systems. Moreover, the adoption of a convention of this nature, as provided for in Article 19 of the WHO Constitution, could be the prelude to reflection on world health governance.

The negotiation and adoption of an international treaty on pharmaceutical R&D was one of the key elements in the implementation of the GSPOA. Indeed, if successful, this could be the most important achievement of the Global Strategy.

2.6.2 The Framework Convention on Tobacco Control

As noted earlier, there is only one historical precedent for the use of Article 19 of the WHO Constitution in one substantive area: the Framework Convention on Tobacco Control (FCTC).

For the first time, the WHO exercised the power to adopt international treaties and agreements in a substantive area and provided a global legal response to a global health threat.

The WHO Framework Convention on Tobacco Control is a framework treaty which, while alluding to many substantive issues, essentially sets out the objectives, principles, institutions, and functioning of what should be a more comprehensive system with the adoption of future additional protocols on technical issues, such as promotion and sponsorship, advertising, illicit trade, and liability (Devillier, 2005, p. 172).

The objective of the Convention is 'to protect present and future generations from the devastating health, social, environmental and economic effects of tobacco consumption and exposure to tobacco smoke'.[8] To this end, the treaty is based on a number of fundamental principles, such as information and protection against the harmful effects of tobacco, multisectoral measures, support for economic conversion, civil society participation, and the principles of partnership and responsibility.

According to the report of the Eighth Session of the Conference of the Parties (COP8) 2018 to the WHO FCTC, Vera Luiza da Costa e Silva, Head of the WHO FCTC, said: 'We are happy to report, based on the information received from the Parties in the 2018 reporting cycle, that progress is evident in implementation of most articles to the Convention, especially the time bound measures concerning smoke-free environments, packaging and labelling and tobacco advertising, promotion and sponsorship ban' (da Costa, 2018).

---

[8] Article 3 of the WHO Framework Convention on Tobacco Control.

2.7   The Evaluation of the GSPOA

Resolution WHA61.21 (2008) establishing the GSPOA also requested the WHO Director-General, among other things, to provide biennial implementation reports in addition to a comprehensive evaluation of the GSPOA after 4 years. In the subsequent resolution WHA62.16 (2009), the Director-General was further requested 'to conduct an overall programme review of the GSPOA in 2014 on its achievements, remaining challenges and recommendations on the way forward to the Health Assembly in 2015 through the Executive Board' (WHO, 2009).

The sixty-eighth session of the World Health Assembly adopted Resolution WHA68.18 (WHO, 2015), in which it decided to extend the time frame of the GSPOA from 2015 to 2022. It further decided to extend the deadline for the overall programme review to 2018. Resolution WHA68.18 agreed to a process for carrying out: (1) a comprehensive evaluation, and (2) an overall programme review. The comprehensive evaluation of the implementation of the GSPOA was to be undertaken by an independent expert evaluator whose work would be overseen by an ad-hoc evaluation management group. The Director-General was also requested to establish a panel of 18 experts to conduct the overall programme review, taking into consideration the findings of the comprehensive evaluation but also other technical and managerial aspects of the programme.

The comprehensive evaluation was intended for documenting 'achievements, remaining challenges and recommendations on the way forward' (WHO, 2015). Thus, the purpose was to assess the status of implementation of the eight elements of the global strategy: (1) prioritising research and development needs, (2) promoting research and development, (3) building and improving innovative capacity, (4) transfer of technology, (5) application and management of intellectual property to contribute to innovation and promote public health, (6) improving delivery and access, (7) promoting sustainable financing mechanisms, and (8) establishing monitoring and reporting systems. The terms of reference of the review that was adopted by the Executive Board in January 2017 included a request for the report of the overall programme review to be presented to the World Health Assembly in 2018 through the 142nd session of the WHO Executive Board.

The overall programme review report was submitted in November 2017. The findings of the review included the following:

1. The fundamental concerns that justified the development of the GSPOA remained valid.
2. R&D is still not sufficiently directed at health products for diseases that mainly affect developing countries, and resources devoted to R&D on these diseases have not sufficiently increased.
3. The 108 action points under the GSPOA are too broad and numerous, which makes it difficult to monitor progress, and stakeholders have devoted very little effort towards implementation of the GSPOA action points.
4. The level of awareness about the GSPOA at the country level is very low.

The review panel found that although the eight elements of the GSPOA were broadly valid, the main problem was the lack of impact in its implementation. The review panel suggested that the review could best add value by recommending a strategy that is more focused in scope and scale and included a set of priority actions for each of the eight elements to address current needs in R&D and access to medicines. Accordingly, the review identified 33 priority action areas, including 17 high-priority actions, with measurable indicators and deliverables. These action areas were identified based on their specificity and feasibility. The WHO and its Member States were specifically responsible for implementation of these action areas. It was recommended that in 2018 the WHO publish a draft implementation plan for these action areas, establish a monitoring mechanism to support implementation, and publish annual reports. Member States were requested to collect and report information to G-Finder.

The Executive Board in January 2017 had also requested the WHO Secretariat to develop an estimate of funding requirements and possible sources for the implementation costs of the recommendations of the programme review. In document EB142/14, the Secretariat estimated that the budget for full implementation of the recommended 33 actions by the review will be $31.5 million over the period 2018–2022. In addition, the budget estimate for the 17 high-priority actions was $16.3 million. According to the Secretariat, this $47.8 million budget would allow the Secretariat to ensure implementation and monitoring of the GSPOA and provide technical support and guidance to Member States in their implementation from 2018 to 2022. The proposed budget exceeded existing resources; therefore, additional resources would need to be mobilised from assessed or voluntary contributions.

The Secretariat has also proposed a draft decision for the consideration of the Executive Board in document EB142/14 Add.1 (2018). The draft decision text requested the WHO Director-General to take forward the recommendations of the review panel following the drawing of a detailed implementation plan in accordance with the review panel's recommendations. Also, the text requested the WHO Director-General to report on progress made in implementing the decision to the World Health Assembly in 2020 through the Executive Board.

The recommendations of the review panel identifying 33 priority action areas, including 17 high-priority action areas across the eight elements of the GSPOA, aimed to provide greater specificity and focus on effective implementation through measurable indicators. WHO Member States endorsed the recommendations of the review panel in the report of the Director-General A71/13.

2.8   The Report of the United Nations Secretary-General's High-Level Panel on Access to Medicines

Towards the end of 2015, at the initiative of UNDP, the Secretary-General of the United Nations convened a High-Level Panel on Access to Medicines (UNHLP). This High-Level Panel published a report of their work on 14 September 2016.

The terms of reference of the UN Secretary-General's call for the High-Level Panel (December 2015) were premised on the existence of a structural problem in the current medical R&D model. Members of the panel were asked to study the '(i) ncoherence between the rights of inventors, international human rights law, trade rules and public health'.

In only 4 months, 180 proposals were received by the High-Level Panel from governments, UN agencies, NGOs, universities, the pharmaceutical industry, and other stakeholders. They can be classified into five categories:

1. Comments on the current R&D model (40)
2. Proposals to strengthen health systems (27)
3. Proposals to modify the R&D model progressively (46)
4. Contributions proposing a major reform of the model (46)
5. Other

Proposals were received from the governments of the Netherlands, Lesotho, Japan, and Jordan.

The main recommendations of the UNHLP report released in September 2016 can be summarised as follows:

- Make use of the available space in TRIPS Article 27 to adapt and apply rigorous definitions of invention and patentability.
- Governments should adopt and implement legislation facilitating compulsory licenses.
- WTO members should review the paragraph 6 decision.
- Governments and the private sector must refrain from explicit or implicit threats, tactics, or strategies that undermine the right to use TRIPS flexibilities.
- No to TRIPS-plus provisions.
- Universities and research institutions receiving public funding should prioritise public health objectives over financial profitability in their patent and licensing practices.
- All interested parties should test and implement new and additional models of research funding (R&D).
- The UN Secretary-General should initiate a process for governments to negotiate global agreements on the coordination, financing, and development of health technologies, including negotiations for a binding R&D Convention to delink the cost of R&D from the final price of medicines, thus promoting access to good health for all. Governments should establish a working group to initiate the negotiation of a Code of Principles for Biomedical R&D.

- Governments should review the status of access to health technologies in their country through the lens of human rights principles.
- Governments should require manufacturers and distributors to disclose to drug regulatory and procurement authorities information on the cost of R&D, production, marketing, and distribution of health technologies.
- Governments should make all clinical trial data publicly available (United Nations, 2016).

Although the discussions leading to the production of the report were not public, dissenting comments by some members of the panel at the end of the report clearly show that consensus was not reached on some of the recommendations, which would have otherwise significantly advanced the debate on the need for making substantive changes to the current R&D model to improve access to medicines.

One of the most significant contributions to the debate on access to medicines made by the UNHLP report is the assertion that this is a global problem that affects both developing and developed countries. All documents produced in the WHO context stated that the problem encompassed some diseases that disproportionately affected developing countries. A report produced after the U.S. commercialisation of Sofosbuvir for Hepatitis C, at a price of $84,000 for a 12-week treatment, could not continue to claim that the problem was only limited to poor countries.

The second most important contribution of the report is the recommendation to 'make full use of the policy space available in Article 27 of the TRIPS Agreement by adopting and applying rigorous definitions of invention and patentability'. This is undoubtedly the most important flexibility of the TRIPS agreement, that is, the freedom of each country to interpret and define the three requirements of the TRIPS agreement to grant patents: novelty, inventiveness (non-obviousness), and industrial application (utility).

The third important point of the report is not new, but it is critical in that it rescues a recommendation that already existed in the context of WHO, but which countries and the WHO Secretariat were unable to put into practice: to begin 'negotiations for a binding R&D Convention that delinks the costs of research from end prices to promote access to good health for all' (United Nations, 2016). In the 180 contributions from countries, institutions, UN agencies, NGOs, universities, the pharmaceutical industry, and individuals from around the world, one-third alluded to some form of treaty or binding convention as an alternative or complement to the current model for R&D primarily based on patent protection.

The fourth important point concerns the almost symbolic contribution that the WTO has made to the problem of access to medicines until now with the so-called 'paragraph 6', a mandate given by the Doha Declaration, which has given no results yet after 13 years of adoption of waivers to Article 31 of the TRIPS Agreement. The report of the UNHLP recommends that WTO members review what is known as the 'paragraph 6 decision' adopted in 2003.

2.9   The Roadmap on Access to Medicines

2.9.1   Background

The 2018 World Health Assembly adopted decision WHA71(8) that requested the WHO Director-General to elaborate a roadmap report on access to medicines and vaccines for the 2019–2023 period and submit that report to the World Health Assembly in 2019 through the 144th session of the Executive Board. Consultations were held on a zero draft of the roadmap with Member States and intergovernmental organisations and non-state actors from July–September 2018. Based on the feedback received from these consultations, the draft roadmap was updated and presented for the consideration of the Executive Board in January 2019. The Executive Board took note of this report. A revised version of this report was presented for the consideration of the World Health Assembly in May 2019. The revision added a new appendix 2 to the document to indicate the linkage between the Thirteenth General Programme of Work, 2019–2023, and the activities, actions, deliverables, and milestones set out in the roadmap. It also reflects issues raised by the Executive Board relating to providing health products for primary health care, monitoring access, optimising the use of biosimilars, addressing the challenges faced by Small Island States, and supporting countries transitioning from donor funding. The 2019 World Health Assembly took note of the draft roadmap report.

The revised roadmap aligns to the following outputs of the WHO General Programme of Work for 2019–2023: (a) providing guidance on quality, safety, and efficacy of health products, including through prequalification services, essential medicines, and diagnostics lists, (b) improved and more equitable access to health products through global market shaping and supporting countries to monitor and ensure efficient and transparent procurement and supply systems, (c) strengthening country and regional regulatory capacity and improving supply of quality-assured and safe health products, (d) defining the R&D agenda and coordinating research in line with public health priorities, and (e) enabling countries to address antimicrobial resistance through strengthened surveillance systems, laboratory capacity, infection prevention and control, awareness-raising, and evidence-based policies and practices.

The roadmap seeks to address two broad strategic objectives: (a) ensuring quality, safety and efficacy of health products, and (b) ensuring equitable access to health products. The roadmap describes activities, specific actions, and deliverables for each of these strategic areas. On quality, safety, and efficacy, the roadmap focuses on regulatory system strengthening, prequalification, and market surveillance. Concerning equitable access, the roadmap focuses on aligning R&D to public health needs, application and management of intellectual property, evidence-based selection, fair and affordable pricing, procurement and supply chain management, appropriate prescribing, dispensing, and rational use.

The roadmap states at the outset that it is based on key World Health Assembly resolutions over the last 10 years relating to access to medicines. This implies that the roadmap considers resolutions that go back up to 2008 only. Hence, it ignores several major World Health Assembly resolutions prior to 2008 that give the WHO a specific mandate for activities on access to medicines and the use of TRIPS flexibilities to that end. These include resolutions WHA49.14, WHA52.19, WHA53.14, WHA54.10, WHA57.14, WHA58.34 and WHA59.26.

The following action areas under the roadmap are ambiguous or do not respond to the GSPOA: regulatory systems strengthening, health research and development, application and management of intellectual property, ensuring fair pricing, and reducing out-of-pocket payments.

2.9.2   Regulatory Systems Strengthening

Concerning regulatory systems strengthening, the roadmap refers to the role of the WHO in developing regulatory norms and standards and expanding reliance on national regulatory authorities that meet international performance benchmarks under the WHO Global Benchmarking Tool for assessment of national regulatory systems. The roadmap focuses on the promotion of work-sharing and convergence among national regulatory systems. This appears to be an implied reference to the promotion of regulatory harmonisation. It should be recalled that in negotiations during the 2014 WHO Assembly on resolution WHA67.20, developing countries had strongly objected to any reference to promotion of regulatory harmonisation or the inclusion of standards developed by the International Conference on Harmonisation (ICH) – a partnership of regulatory agencies of developed countries and multinational pharmaceutical companies to which the WHO is a permanent observer. In this context, it will be critical to ensure that WHO activities in the area of regulatory systems strengthening are not unduly influenced by commercial interests of multinational pharmaceutical companies and lead to harmonisation of untenable regulatory standards for developing countries.

2.9.3   Health Research and Development

On health research and development, the roadmap does not go beyond the business-as-usual approach and limits itself to gathering and processing information under the Global Observatory on Health Research and Development. There is no mention of the recommendations of the Consultative Expert Working Group on Research and Development: Financing and Coordination (CEWG) for negotiating a global biomedical R&D treaty, the need for which has also been endorsed by the report of the United Nations Secretary-General's High-Level Panel on Access to Medicines (UNHLP).

2.9.4 Intellectual Property

On intellectual property (IP), the roadmap focuses on the application of appropriate IP rules and management of IP for fostering innovation and access to health products and providing technical support and capacity building. On the application and management of IP rules, the roadmap focuses on promotion of public health oriented licensing agreements, transparency on patent status of health technologies, sharing country experiences on public health approaches to the use of TRIPS flexibilities, review of mechanisms and initiatives for access to affordable health technologies enabled by publicly funded R&D, and support for the expansion of the Medicines Patent Pool to patented essential medicines in the WHO treatment guidelines through identification of potential products for licensing. The WHO can also provide on-demand technical assistance to countries on making use of the TRIPS flexibilities, assessing the public health implications when negotiating bilateral and multilateral trade agreements, assessment of the patent status of essential medicines. The roadmap also focuses on the continuation of the trilateral cooperation with the WIPO and WTO, and also with UNCTAD and UNDP.

While it is important that WHO provides support to countries in adopting a public health approach to the use of TRIPS flexibilities, it will be essential to ensure that WHO raises awareness about the importance and the full scope of the TRIPS flexibilities for access to medicines. However, the roadmap does not make any mention of the importance and scope of TRIPS flexibilities in the introduction of the action areas in the report. Hence, while technical support for the use of TRIPS flexibilities is WHO, it is somewhat undersold in this report. Another aspect of the roadmap is that it focuses on management and licensing of IP rights which is not within the competence of WHO. The report also gives undue prominence to the trilateral collaboration between WHO-WTO-WIPO.

The roadmap also refers to ensuring fair pricing as an action area. In this regard, it is important to stress that there is no common understanding of fair pricing among WHO Member States.

2.10 Resolution on 'Improving the Transparency of Markets for Medicines, Vaccines and Other Health-Related Technologies'

Italy, Greece, Malaysia, Portugal, Serbia, Slovenia, South Africa, Spain, Turkey and Uganda proposed a resolution for adoption by the 2019 World Health Assembly. The resolution aimed to improve transparency around prices of medicines, vaccines and other health technologies. The resolution was presented in the context of the roadmap for access to medicines, on promoting transparency in the prices of medicines, vaccines and health technologies. The draft resolution expressed concern about high prices of medicines, vaccines and health technologies and that these high prices could impede progress towards achieving Universal Health Coverage (UHC). To that end, the draft resolution sought to enhance publicly available information on costs of manufacturing medicines, vaccines and health technologies and the patent landscape of medical technologies. The draft resolution also expressed concern about the limited public availability of complete and comprehensive data on clinical trials. The draft resolution further stated that availability of reliable, transparent and sufficiently detailed data on the costs of R&D inputs, medical benefits and therapeutic value of a product could facilitate better evaluation of policies that influence pricing of health technologies or appropriate rewards for research outcomes.

Thus, the draft resolution urged Member States of WHO to:

- undertake measures to publicly share information on prices and reimbursement costs of medicines, vaccines, cell and gene-based therapies, and other health technologies
- require that all human subject clinical trial results be reported publicly, including the costs incurred to undertake each clinical trial and the direct funding, tax credits and other subsidies received from governments
- require submission of annual reports on sales revenues, prices and units sold, annual reports on marketing costs for each registered product or procedure, R&D costs directly associated with clinical trials, grants, tax credits and public sector subsidies or incentives relating to the initial regulatory approval
- improve the transparency of the patent landscape of medical technologies, using approaches that do not create barriers to generic competition through sharing complete and up-to-date information

The draft resolution also requested the WHO Secretariat to:

- support Member States in collecting, analysing and creating standards for information on prices, reimbursement costs, clinical trials outcome data, and costs of relevant policy development and implementation towards UHC
- create a web-based tool for governments to share information on medicine prices, revenues, R&D costs, public sector investment and subsidies for R&D, marketing costs, and other related information
- create an experts' forum to develop suitable options for alternative incentive frameworks to 'patent monopolies' for new medicines and vaccines
- create a biennial forum on the transparency of markets for medicines, vaccines and diagnostics to evaluate progress towards expansion of transparency; and to report to the 146th Executive Board on the measures that are needed for the WHO Global Observatory on Health R&D to enhance reporting on pre-clinical investments in R&D both by public and private sectors

It is indeed very important to ensure transparency on the cost factors that contribute to the price of a medicine, vaccine or any other health technology to develop appropriate policy responses to ensure affordable pricing. Unfortunately, pharmaceutical companies do not readily make this information available and have resisted suggestions to do so. However, nothing prevents any WHO Member State from adopting regulations requiring pharmaceutical companies to disclose such information. If required, any WHO Member State can also seek the assistance of WHO in this regard.

This resolution, which is supposed to be the continuation or part of the implementation of the GSPOA, ignores the central points of the GSPOA that refer to intellectual property.

The resolution lacks reference to the work already undertaken in WHO such as the recommendations of the Consultative Expert Working Group on Research and Development (CEWG) which recommended, among others, the adoption of measures to promote transparency under a binding R&D treaty. Indeed, many elements of the CEWG recommendations relating to transparency are better reflected in the CEWG report. For example, on human clinical trials, the focus of the resolution is on transparency on the cost components of the clinical trials. Although the WHO Secretariat is requested to support Member States on clinical trials outcome data, the CEWG recommendations in fact clearly called for transparency through standards on disclosure of information on the appropriateness of specific clinical trials and the benefits of the same, and not just the cost of such trials. Further, the CEWG recommendation also called for transparency in terms of licensing agreements relating to R&D outcomes.

Unlike the initial draft, first proposed by Italy and then supported by a group of countries, most of them industrialised countries, the resolution does not create any responsibility on the part of WHO or Member States to ensure transparency of the cost of pharmaceutical R&D and clinical trials.

In addition, the resolution urges Member States to take measures to disclose the price of medicines, an activity that WHO has been doing for more than 10 years and for which a new mandate was not necessary.

The resolution clearly highlights the division within Europe between countries where the pharmaceutical industry is strong, such as Germany, France, the United Kingdom, Switzerland, Sweden and Denmark, on the one hand, and countries with a less important pharmaceutical industry such as Spain, Portugal, the Netherlands, Austria and Norway, on the other (Gopakumar, 2019).

Although the resolution was adopted by 'consensus', Germany, Hungary and the United Kingdom declared their dissociation from the resolution in the plenary. The United Kingdom insisted on the fact that the resolution was not submitted first, as it is customary to the Executive Board and also announced their disassociation from the resolution. The United States supported the final version of the resolution indicating that, after the reforms introduced at the demand of the Europeans, the final text was not inconvenient for the industry. Spain said it would prefer fewer reservations and more provisions on the costs of R&D of pharmaceuticals. France confirmed, as it had already expressed from the outset, its opposition to the resolution.

2.11 Access to Biotherapeutic Products Including Similar Biotherapeutic Products

Another critical issue for Member States in the context of the GSPOA is the implementation of the WHA resolution 67.21 from 2014 on 'Access to biotherapeutic products including similar biotherapeutic products and ensuring their quality, safety and efficacy'. The resolution had requested the WHO Expert Committee on Biological Standardization to update the 2009 guidelines, taking into account the technological advances for the characterisation of biotherapeutic products and considering national regulatory needs and capacities and to report on the update to the Executive Board. However, to date, the WHO Secretariat has not updated the Similar Biotherapeutic Products (SBP) guidelines. The Secretariat has reported that after the adoption of the WHA resolution '(i)n April 2015, an informal consultation was organized during which participants from National Regulatory Authorities of both developing and developed countries, as well as from industry, recognized and agreed that the evaluation principles described in the Guidelines were still valid, valuable and applicable in facilitating the harmonisation of SBP regulatory requirements globally. It was therefore concluded that there was no need to revise the main body of the existing Guidelines' (WHO, 2018).

This is a problematic approach as the referred-to resolution clearly requested the Director-General to convene a meeting of the WHO Expert Committee on Biological Standardization to update the Guidelines. The resolution did not leave it to the discretion of the Expert Committee on Biological Standardization to decide whether to update the guidelines or not. Further, a decision or resolution by Member States in the World Health Assembly cannot be overturned by an informal consultation. The WHO has not published any verbatim record or minutes of the 2015 informal meeting.

2.12 Conclusions

The IGWG negotiation is undoubtedly the most important exercise ever carried out by WHO Member States in relation to access to medicines; and it was an exceptional opportunity for the WHO to exercise its leadership by proposing a vision and mechanisms for the following 15 to 20 years. This negotiation which went on for 2 years, can be considered the most relevant and important in the almost 70 years of existence of the WHO, second only to the negotiation and adoption of the convention against tobacco, the Framework Convention on Tobacco Control (FCTC).

Did the WHO Secretariat have a vision and clarity regarding the direction of the strategy, and enough independence to accompany the countries' efforts? This was the fundamental question for which we unfortunately still do not have a clear answer, 10 years later.

There is no denying that progress over the last 10 years has been enormous. The issue of the impact of intellectual property on access to medicines has fully entered into the debate on access to health and today also into the debate on universal health coverage (UHC). It is impossible to think about UHC without universal access to medicines.

As was seen at the last World Health Assembly (2019), in the discussions of the failed resolution called the transparency resolution, delegates from developing countries have a clear knowledge of the issue. A small group of industrialised countries, where the large pharmaceutical industry is located, continues to oppose what was and remains the heart of the GSPOA: the adoption of 'a binding international treaty for R&D and innovation for health', as permitted by Article 19 of the WHO Constitution and as recommended by the different WHO reports and WHA Resolutions. This point today seems far from the interest of the WHO Secretariat or of the countries that promoted it simply because of the lack of space to continue discussing such a politically sensitive issue. If this is the case, the need for new ideas and models on how to prioritise, organise and finance R&D for medicines remains valid.

The GSPOA sought a substantial reform of the pharmaceutical research and development system in view of this system's failure to produce affordable medicines for diseases affecting the majority of the world's population living in developing countries. The intellectual property rights required by the TRIPS Agreement and recent trade agreements have become obstacles to access to medicines. The GSPOA made a critical analysis of this reality and opened the door to the question of new solutions to this problem (Velásquez, 2011).

References

Abbott, F. M., & Reichman, J. H. Strategies for the protection and promotion of public health arising out of the WTO TRIPS agreement amendment process. Florida State University and Duke University.

Commission on Intellectual Property Rights. (2002). Integrating intellectual property rights and development policy: Report of the Commission on Intellectual Property Rights, Executive summary, p. 14. London, September 2002.

Correa, C. M. (2004, October). Ownership of knowledge – The role of patents in pharmaceutical R&D. *Bulletin of the World Health Organization, 82*(10), 719–810.

da Costa, V. L. (2018). Opening remarks at the COP8. Geneva, 1 October 2018. https://www.who.int/fctc/secretariat/head/statements/2018/cop8-open-remarks-head-secretariat/en/.

Devillier, N. (2005). La convention-cadre pour la lutte anti-tabac. *Revue Belge du Droit International, 1-2,* 172.

Gopakumar, K. M. (2019). WHO: Member States adopt resolution on transparency in medicine pricing. TWN Services on Health Issues, May 2019.

KEI. (2009). WHA: Civil Society letter to WHO Member States. 18 May 2009. https://www.keionline.org/21002.

Pedrique, B., Strub-Wourgaft, N., Some, C., Olliaro, P., Trouiller, P., et al. (2013). The drug and vaccine landscape for neglected diseases (2000-11) a systematic assessment. *The Lancet Global Health, 1*(6), E371–E379. https://doi.org/10.1016/S2214-109X(13)70078-0

Rottingen, J. A. (2011). PPT presentation, Geneva, July 2011.

Seuba, X. (2008). *La protección de la Salud ante la regulación internacional de los productos farmacéuticos*. Doctoral thesis p. 92 and ff., Barcelona.

Smith, R. D., Correa, C. M., & Oh, C. (2009). Trade, TRIPS and pharmaceuticals. *The Lancet, 373*, 687.

Trouiller, P., et al. (2002). Drug development for neglected diseases: A deficient market and a public health policy failure. *The Lancet, 359*, 2188.

UNAIDS. (2012). Global report: UNAIDS report on the global AIDS epidemic 2012. https://www.unaids.org/sites/default/files/media_asset/20121120_UNAIDS_Global_Report_2012_with_annexes_en_1.pdf.

UNAIDS. (2019). UNAIDS data 2019. https://www.unaids.org/en/resources/documents/2019/2019-UNAIDS-data.

United Nations. (2016). The United Nations Secretary-General's High-Level Panel on Access to Medicines Report: Promoting innovation and access to health technologies. 14 September 2016. http://www.unsgaccessmeds.org/final-report.

Velásquez, G. (2011). *The right to health and medicines: The case of recent negotiations on the global strategy on public health, innovation and intellectual property*. Research Paper No. 35. : South Centre, January 2011. https://www.southcentre.int/wp-content/uploads/2013/05/RP35_Right-to-health-and-medicines_EN.pdf.

WHO. (2003a). WHO framework convention on tobacco control. https://www.who.int/fctc/text_download/en/.

WHO. (2003b). World Health Assembly, "Intellectual Property Rights, Innovation and Public Health", WHA Resolution 56.27, 28 May 2003, para. 2.

WHO. (2006a). Constitution of the World Health Organization, basic documents, 45th edn, Supplement, October 2006. https://www.who.int/governance/eb/who_constitution_en.pdf.

WHO. (2006b). Public health, innovation and intellectual property rights. Report of the Commission on Intellectual Property Rights, Innovation and Public Health (CIPIH). https://www.who.int/intellectualproperty/documents/thereport/ENPublicHealthReport.pdf?ua=1.

WHO. (2007a). Report by the Secretariat, 31 July 2007, para. 11.

WHO. (2007b). Report of First Session, 25 January 2007, paras. 20, 21 and 31.

WHO. (2008). Sixty-First World Health Assembly. WHA61/2008/REC/1. Geneva, 19–24 May 2008. Resolutions and decisions annexes. https://apps.who.int/gb/ebwha/pdf_files/WHA61-REC1/A61_REC1-en.pdf.

WHO. (2009). Sixty-Second World Health Assembly. WHA62/2009/REC/1. Geneva, 18–22 May 2009. Resolutions and decisions annexes. https://apps.who.int/gb/ebwha/pdf_files/WHA62-REC1/WHA62_REC1-en.pdf.

WHO. (2015). WHA68.18. Agenda item 17.5. Global strategy and plan of action on public health, innovation and intellectual property. 26 May 2015. https://apps.who.int/gb/ebwha/pdf_files/WHA68/A68_R18-en.pdf?ua=1.

WHO. (2018). WHO questions and answers: Similar biotherapeutic products. Complementary document to the WHO Guidelines on evaluation of similar biotherapeutic products (SBPs). https://www.who.int/biologicals/expert_committee/QA_for_SBPs_ECBS_2018.pdf?ua=1.

WHO. (2019). Roadmap for access to medicines, vaccines and other health products, 2019–2023. https://apps.who.int/gb/ebwha/pdf_files/WHA72/A72_17-en.pdf.

Wibulpolprasert, S., et al. (2007). WHO's web-based public hearings: Hijacked by pharma? *The Lancet, 370*(24), 1754.

World Health Assembly, 65. (2012). Consultative expert working group on research and development: Financing and coordination. WHO. https://apps.who.int/iris/handle/10665/79197.

World Vision. (2020). Global poverty: Facts, FAQs, and how to help. https://www.worldvision.org/sponsorship-news-stories/global-poverty-facts#:~:text=1990%3A%20The%20World%20Bank%20defined,less%20than%20%20%241.25%20a%20day.

WTO. (2001). Doha WTO Ministerial 2001: TRIPS, WT/MIN(01)/DEC/2, 20 November 2001 Declaration on the TRIPS agreement and public health, Adopted on 14 November 2001. https://www.wto.org/english/thewto_e/minist_e/min01_e/mindecl_trips_e.htm.

WTO, WIPO, WHO. (2012). Promoting access to medical technologies and innovation intersections between public health, intellectual property and trade. https://www.wto.org/english/res_e/booksp_e/pamtiwhowipowtoweb13_e.pdf.

The opinions expressed in this chapter are those of the author(s) and do not necessarily reflect the views of the SC: South Centre, its Board of Directors, or the countries they represent.

Chapter 3
Rethinking Global and Local Manufacturing of Medical Products After COVID-19

3.1 Introduction

The objective of this document is to examine how the great challenge caused by COVID-19 in 2020 in the area of production of medicines and health products can be used as an opportunity to improve and strengthen access to medicines in developing countries: 'Major crises bring about challenges but also opportunities. The strategic importance of a local pharmaceutical industry has been increasingly recognised as a result of the COVID-19 crisis. Developing countries should take advantage of this opportunity to strengthen their pharmaceutical industry, including biological medicines' (Correa, 2020).

In Sect. 3.2 of this chapter (Background: The View of UN Agencies on Pharmaceutical Production in Developing Countries), the role of the United Nations (UN) agencies in the last 30 years is analysed in relation to the local production[1] of medicines. As examined there, although the United Nations Industrial Development Organization (UNIDO) and United Nations Conference on Trade and Development (UNCTAD) have tried to promote and support the local production of medicines, agencies such as the World Health Organization (WHO) have not been clear or have even advised against local production in developing countries.

In Sect. 3.3 (COVID-19 'Vaccine Nationalism'), the document analyses the trends originated by the new realities that the health crisis has brought to light, notably the interdependence in terms of pharmaceutical production and the phenomenon that has been termed 'vaccine nationalism'. This section also refers to the massive

---

[1] For this chapter, 'local production' refers to manufacturing of pharmaceuticals by local state-owned pharmaceutical companies, local private pharmaceutical companies, and joint-ventures of local private or state-owned and foreign pharmaceutical companies.

This chapter is largely taken from: Velásquez, G. (2020 September). *Re-thinking Global and Local Manufacturing of Medical Products After COVID-19*. South Centre Research Paper 118. https://www.southcentre.int/wp-content/uploads/2020/09/RP_118_reduced-1.pdf.

© SC: South Centre 2022
G. Velásquez, *Vaccines, Medicines and COVID-19*, SpringerBriefs in Public Health, https://doi.org/10.1007/978-3-030-89125-1_3

public subsidies to the private sector in some developed countries, without sufficiently clear rules and conditions.

Section 3.4 (COVID-19 Global Vaccine Access Facility) analyses the role of the new initiative, the COVAX Facility, its shortcomings, and the concerns of some NGOs about the absence of conditions that should ideally accompany the unprecedented financial subsidies that have been largely granted with public funds.

Section 3.5 (Global Preparedness Monitoring Board) shows that COVID-19 could not be regarded as a total surprise, something unexpected – we had already been warned. In May 2011, a WHO document on pandemic influenza preparedness alerted countries to the continuing risk of an influenza pandemic with potentially devastating health, economic and social consequences, particularly for developing countries, which have a higher disease burden and are more vulnerable (WHO, 2011b).

Section 3.6 (A COVID-19 Technology Sharing Platform: A Recent UN initiative) addresses a recent (May 2020) initiative by three UN agencies, including the WHO, to support access to technology for the local production of medicines and health products. It would seem that the challenge of COVID-19 has led the UN agencies to seek mechanism to improve access to technologies and thereby to medicines and other health products in developing countries.

The chapter concludes by noting that a reorganisation of global pharmaceutical production could perhaps be beneficial to increasing access to medicines in developing countries, and states (public sector) should be more involved in the promotion of the production of essential inputs for health systems. This could be an opportunity to ensure that health, rather than purely commercial gains, becomes the main objective of the pharmaceutical industry.

Finally, this chapter does not refer to the necessary investments, technologies, scales of production, competitiveness, etc., important aspects when talking about local production. The main objective of this chapter is to reopen the debate on an issue that had been somehow left aside and that now regains urgent relevance with the COVID-19 crisis.

3.2   Background: The View of UN Agencies on Pharmaceutical Production in Developing Countries

The unprecedented global health crisis caused by the coronavirus disease (COVID-19) pandemic during the first quarter of 2020 has reopened the discussion about local pharmaceutical production, which has become now particularly relevant and urgent. The COVID-19 crisis has highlighted the global interdependence in the supply of pharmaceuticals. No country is self-sufficient. Many industrialised countries are taking the decision to repatriate or develop the production of active pharmaceutical ingredients (APIs). Many governments are beginning to talk about pharmaceutical sovereignty and/or health security (Correa, 2020). If this becomes a reality, developing countries will have to begin and/or strengthen local production of medicines and vaccines (Syam, 2020). In particular, the war to obtain the future vaccine for COVID-19 does not look easy with these new developments as they may further concentrate the control over vaccines' production in a few developed countries. Currently, about 80% of global vaccine sales come from five large multinational corporations.[2]

As early as the 1980s, three agencies in the United Nations system were already interested in the local manufacturing of drugs in developing countries: UNIDO (UNIDO, 1980) and UNCTAD, which provided technical assistance in the area of the transfer of technology in the pharmaceutical field, (Stork & Wanandi, 1980, p. 72.) and the WHO, which created the Action Programme on Essential Drugs (Velásquez, 1986).

During its first 20 years, the WHO Action Programme on Essential Drugs gave priority to the development of national drug policies, and its position on drug manufacturing in developing countries was always ambiguous or openly contrary to it. Thus, Kapan and Laing stated in 2005 that 'if a developing country with manufacturing facilities is able to finish off bulk active ingredients sourced from developed or other countries at high costs, such manufacture may have no impact whatever on patient access to needed medicines' (Kaplan & Laing, 2005).

It is clear from the findings of Kaplan and Laing, (the latter was responsible for this area in the WHO Medicines Programme) that the WHO was not, at that time, in favour of promoting the production of medicines in developing countries:

'[O]ur preliminary conclusions are:

- In many parts of the world, there is no reason to produce medicines domestically since it makes little economic sense.
- In the local pharmaceutical manufacturing sector, local production is often not reliable and, even if reliable, it does not necessarily mean that medicine prices are reduced for the end user.
- If many countries adopt local production, the result may be less access to medicines since production facilities in many countries may mean forgoing economies of scale.
- It may be possible for small country markets to be coordinated or otherwise joined together to create economies of scale. (…)
- For many countries, technical expertise, raw materials, quality standards, and production and laboratory equipment need to be imported, with the result that foreign exchange savings may be small or non-existent.
- Few developing countries have the capacity to produce active ingredients for pharmaceutical manufacture' (Kaplan & Laing, 2005).

A WHO literature review of local production and access to medicines in low- and middle-income countries published in 2011 concludes:

---

[2] See WHO, Vaccine market. Global Vaccine Supply, https://www.who.int/immunization/programmes_systems/procurement/market/global_supply/en/.

- 'We note the predominance of case studies and surveys and the relative lack of econometric and time series studies linking local production and access.
- Our brief review of the UNCTAD technology transfer literature does not suggest any robust attempt to link local production and access to medicines, but this may not be surprising, as technology transfer may be considered industrial rather than health policy, and the case study methodology is not strictly applicable to investigate such a link.
- The business and economic literature that we have seen also is concentrated on the upstream side (e.g., supply side, industrial policy, knowledge spill-overs, innovation etc.) with seemingly little information on the downstream issues of local production and access to medicines.
- The public health literature on the subject of local production is directed predominantly towards the issue of intellectual property rights and access to medicines.
- It seems quite remarkable that many of the pricing surveys do not distinguish the price of local versus foreign producers on a product-by-product basis.
- There is an almost complete absence of information on the link between local production and access to medical devices (…)' (WHO, 2011a).

Local production has been a subject of discussion in the World Health Assembly (WHA) since the 1970s. Element 4 of the resolution 61.21 (2008) on a Global Strategy and Plan of Action on Public Health, Innovation and Intellectual Property (GSPOA) is about the promotion and transfer of technology, and the production of health products in developing countries is the first recommended action (WHO, 2008).

In this context, and as recommended by the GSPOA, a project to explore ways in which local production and technology transfer could be strengthened in a number of low- and middle-income countries was launched by the WHO in 2009 in cooperation with UNCTAD. The project, titled 'Improving access to medicines in developing countries through technology transfer related to medical products and local production', concluded in September 2016.

Several UNCTAD publications, in the context of this project, analyse and promote the local production of medicines in developing countries.

In the document 'Tool Box for Policy Coherence in Access to Medicines and Local Pharmaceutical Production', for instance, UNCTAD presents s an overview of policy tools that developing countries may consider to create a framework conducive for promoting local pharmaceutical production and access to medicines: 'As the promotion of local pharmaceutical production depends on the coordination of various areas of policy, such as drug regulation, research and development (R&D), investment, trade and intellectual property, the Tool Box emphasises the importance of ensuring coherence among policies that at first sight appear unrelated to each other' (UNCTAD, 2017).

In another publication, 'Local Production of Pharmaceuticals and Related Technology Transfer in Developing Countries', UNCTAD analyses several case studies from Argentina, Bangladesh, Colombia, Ethiopia, Indonesia, Jordan Thailand and Uganda (UNCTAD, 2011).

By giving concrete examples of successful technology transfer initiatives in the area of pharmaceutical production, the UNCTAD case studies 'provide a number of important lessons for policy-makers and other stakeholders in both developing and developed countries on issues of investment, science, technology and innovation, and intellectual property rights' (UNCTAD, 2011).

In April 2017, the WHO convened an interagency consultation to discuss local production of essential medicines. The meeting was held in Geneva and was attended by representatives of 14 international agencies. Given the position that the WHO took on the issue, not surprisingly, one of the conclusions of the consultation was: 'While it may be feasible to develop local production initially, commercial sustainability remains a challenge when the medicines and health products produced through local production can be more expensive than the commercially available alternatives including imported products' (WHO, 2017).

Interestingly, another UN agency, UNIDO, has held a position quite different from that dominating in the WHO. Since 2006, UNIDO has provided technical cooperation and advisory services to advance local pharmaceutical production (LPP) in developing countries with a wide range of partners. Under a global project, UNIDO contributed to improving the operational environment and technical capacities of local manufacturers and helped 'mainstream' LPP as a global development theme. This engagement has established UNIDO as a leading organisation on the LPP agenda. For UNIDO, LPP is important for developing countries for several reasons:

- 'More than two billion people worldwide cannot get the medicines they need.
- LPP can help vulnerable populations, especially those in remote rural areas, to access quality medicines, thus contributing to "leaving no one behind, and reaching the furthest behind first", the overarching principle of the 2030 Agenda for Sustainable Development.
- LPP can reduce dependency on international donations and the shrinking number of overseas companies that dominate the global market.
- LPP is easier to control and can help curb the vast influx of substandard medicines into developing countries' (UNIDO, 2020).

All the above seems to indicate that the position of the WHO in contrast to other UN agencies, such as UNCTAD and UNIDO, has been that if the production capacity of developed countries is sufficient to supply the world market, it is not worth promoting the production of medicines in developing countries. As we will see below, this assumption is challenged by the nationalism in the production of medicines and vaccines that seems to have emerged with COVID-19.

3.3   COVID-19 'Vaccine Nationalism'

As noted, one of the realities that the health crisis caused by COVID-19 has made evident is the interdependence between all countries in terms of the production of medicines and vaccines. The pandemic has shown, for example, the extent to which developed countries depend on inputs from countries outside the United States (US) and the European Union (EU), notably from China and India.

Beyond the fights over drug markets or future vaccines related to the pandemic, the United States of America, Germany, France and the United Kingdom are now considering how to recover their pharmaceutical sovereignty to depend less on other countries (Lopez Girondo, 2020).

The European Commission launched an online public consultation on the Pharmaceutical Strategy for Europe. Coming in the wake of the COVID-19 pandemic, the Strategy, which will also inform the newly proposed EU4Health Programme and align with Horizon Europe for research and innovation, will aim to ensure Europe's supply of safe and affordable medicines to meet patients' needs and support the European pharmaceutical industry in remaining an innovator and world leader (European Commission, 2020a).

Stella Kyriakides, European Commissioner for Health and Food Safety said: 'The Pharmaceutical Strategy for Europe is a cornerstone of our policy in the area of health for the next five years, (…) we will be responding to the challenges amplified by the COVID-19 pandemic and all the structural issues on access, affordability and the strategic *autonomy* of our Union on medicines' (European Commission, 2020b).

The United States, the European Union and the United Kingdom have purchased the first 2.6 billion doses of vaccines currently in development. The United States Government has given more than $11 billion to eight pharmaceutical companies in the 'Operation Warp Speed' mostly for the development and manufacture of vaccines and has purchased more than 1.2 billion doses. By pre-purchasing doses from the most promising competitors in such large quantities, countries are hedging their bets on which vaccines will be approved first, and how many doses their immunity may require. The industrialised world will be supplied first, and the vaccine will take months or years to reach developing countries (Moore, 2020).

At the time the novel coronavirus started to spread in 2020, it was clear that the stocks or production capacity of masks or alcohol-based hand rub, breathing assistance devices or even the global capacity to produce vaccines, were unknown. Who were the producers and how could they respond to the quantities needed? Prices shot up and some countries imposed export restrictions (Velásquez, 2020). The European Union moved to limit exports of medical equipment outside the EU: 'We need to protect our health workers, who are in the first line of defence against the virus', said Ursula von der Leyen on 15 March 2020 (Bayer et al., 2020).

EU wants to recover the production of medicines 'exiled' in Asia due to low labour costs in that continent. This would be the beginning of deglobalisation in the pharmaceutical sector (Lopez Girondo, 2020).

The German Federal Minister of Health, Jens Spahn, announced his intention to initiate consultations with EU partners about the possibility to relocate the production of certain active pharmaceutical ingredients (APIs) back to Germany (GTAI Germany Trade & Invest, 2020).

He is not the only one worried about pharma supply chains. Emmanuel Macron wants to relocate certain drug production to France. 'The coronavirus pandemic has put the spotlight on health security issues (…) From Thursday, we will launch an initiative to relocate certain critical production', announced the French president (Capital, 2020, author translation). He recently referred to the relocation of pharmaceutical production as a matter of 'pharmaceutical security and industrial sovereignty' (Bezat, 2020, author translation).

By relocating production, industrialised countries have shown they are willing to pay more to protect their pharmaceutical autonomy. Paradoxically, as mentioned above, the WHO has largely discouraged developing countries from producing medicines locally, arguing that locally manufactured products could cost more than imported ones and that the sole priority was to ensure access to low-cost pharmaceuticals.

If the US and EU decide to relocate their pharmaceutical industries and become autonomous in their production of pharmaceuticals, including active ingredients, this would be an opportunity for many developing countries to start or strengthen local production (formulation) of medicines through the import of APIs from China and India as well as to develop, at the national or regional level, their capacity to move up in the value chain and growingly produce APIs. The production of biosimilars offers an opportunity that developing countries should seize, as biologicals account for a growing share of the pharmaceutical market (Lavarello et al., 2018).

A reorganisation of global pharmaceutical production could perhaps be beneficial to increasing access to medicines and other pharmaceuticals in developing countries, and states (public sector) should be more involved in promoting the production of essential inputs for health systems. As states become more involved in the production of medicines and other health products, this could be an opportunity to emphasise and put health objectives ahead of commercial interests. This could be the occasion to make health, rather than purely commercial gains, the main objective of the pharmaceutical industry.

According to *Nature*, as of 7 September 2020, there are more than 231 candidate vaccines against COVID-19 and 33 vaccine candidates are in clinical trials, that are being developed in different parts of the world (Thanh Le et al., 2020). The current COVID-19 vaccine pipeline comprises a broad range of technology platforms, including traditional and novel approaches. Attempts by some governments, such as that of the United States of America, to buy the future vaccine have led to talk of 'vaccine nationalism'. The expression 'vaccine nationalism' describes the circumstance when a country manages to secure doses of vaccine for its own citizens or residents before they are made available to other countries (Santos Rutschman,

2020).[3] This can be done, for instance, through advance market commitments or pre-purchase agreements between a government and a vaccine manufacturer. For instance, in April, the CEO of the French company Sanofi, whose COVID-19 vaccine work has received partial funding from the United States Biomedical Advanced Research and Development Authority (BARDA) announced that the USA had the 'right to the largest pre-order' of its future vaccine (Le Monde avec AFP, 2020, author translation).

The UK Government announced on 12 August 2020 two new agreements that would secure an additional 90 million coronavirus vaccines for its citizens. The in-principle agreements with Novavax and Johnson and Johnson's Janssen bring the UK total number of advance arrangements for a coronavirus vaccine to six, involving four different types of vaccines. Novavax is slated to sell the UK 60 million doses of its candidate, with some to be manufactured in the UK by Fujifilm Diosynth Biotechnologies. The UK will support a Phase 3 clinical trial, with the National Institute for Health Research making its network and expertise available. Janssen would provide 30 million doses of its candidate, which is based on the formula from its successful Ebola vaccine, on a not-for-profit basis. The UK agreed to help pay for global clinical trials of the two-dose immunisation (UK Government, 2020).

In August 2020, Russian President Vladimir Putin stated that a COVID-19 vaccine, dubbed 'Sputnik V' and developed by Russia's Gamaleya Research Institute, had been green-lit for use in the country. The vaccine is being produced primarily for the domestic market, but Moscow is already in talks about exports, the health ministry said. Campaigns of mass vaccinations could start in Russia in December or January (Kayali, 2020).

On 12 August 2020, the US Government announced the purchase of 100 million doses of Moderna's experimental coronavirus vaccine for about $1.5 billion, the Department of Health and Human Services said. The deal gives the government an option to buy another 400 million doses. The US Government has now committed up to $2.48 billion to Moderna's vaccine – including support for late-stage clinical trials, expanded manufacturing and other development activities along with the latest purchase (Brennan, 2020).

Vaccine nationalism is not new. In 2009, during the influenza A (H1N1) pandemic, a similar 'nationalism' arose. Access to vaccines and treatments was determined by purchasing power, and the high-income countries secured the supplies for their populations before the rest of the world (Weintraub et al., 2020). Most of the vaccines for influenza A (H1N1) were bought and stored by the USA, Germany, Belgium, Spain, France, Italy, the Netherlands and Switzerland (Velásquez, 2015). Many developing countries never received their orders, which were placed at the same time as the industrialised countries made their purchase.[4]

---

[3] This expression was used by the WHO Director-General Tedros Adhanom Ghebreyesus at a panel discussion on 6 August 2020. See https://www.swissinfo.ch/eng/reuters/global-recovery-could-be-faster-if-covid-vaccine-made-available-to-all%2D%2D-who-chief/45951960.

[4] Personal communication with Argentine Minister of Health Ginés Gonzalez Garcia.

At that time, several industrialised countries entered into pre-purchase agreements with some vaccine manufacturers. It was said that the global production capacity was 2 billion doses, of which the United States pre-purchased 600 million; All the pre-purchases came from developed countries (Santos Rutschman, 2020).

Some of the world's richest countries fought to be the first to get the vaccines and treatments. Developing countries – among the worst affected – were pushed to the back of the queue, as Western nations signed deals with pharmaceutical producers to guarantee access to vaccines. Australia even stopped a domestic producer from exporting doses to the US until it had immunised its entire population. For many global health experts, the swine flu was a warning for the far more serious coronavirus crisis, which has already killed more than 800,000 people as of 26 August 2020 and brought economies around the world to a standstill. The current COVID-19 pandemic could lead to a geopolitical fight over vaccines that would exceed the one that occurred over the influenza A (H1N1) pandemic (Milne & Vrow, 2020). It has been rightly noted in this regard that:

> For those who believe that a vaccine for COVID-19 will end or largely contain this pandemic or who hope that new drugs will be discovered to combat its effects, there is plenty cause for concern. Instead of working together to craft and implement a global strategy, a growing number of countries are taking a 'my nation first' approach to developing and distributing potential vaccines or other pharmaceutical treatments (Weintraub et al., 2020).

The result of this vaccine nationalism will be that the vaccine may take months, if not years, to reach most developing countries. Perhaps, as in the past, companies or countries will make a symbolic donation of their vaccines to poor countries through the WHO (WHO, 2007). This will not be a sustainable solution.

This approach towards moving away from a collective, global and equitable strategy to confront and combat the pandemic is exemplified by several recent events and statements:

- Access to COVID-19 Tools (ACT) Accelerator: on 4 May 2020, the EU 'Commission registered €7.4 billion, equivalent to $8 US billion, in pledges from donors worldwide during the Coronavirus Global Response pledging event' (European Commission, 2020a). Leaders said that each euro or dollar will be channelled through global health organisations such as CEPI,[5] Gavi, the Vaccines Alliance, and the Global Fund and Unitaid (Wintour, 2020). CEPI and Gavi will work under the umbrella of the Access to COVID-19 Tools (ACT) Accelerator Vaccine Taskforce (Gavi, 2020c). Who will be the partners in this ACT Accelerator initiative? Public sector, industry, research, funders, regulators and international organisations. 'Business partners will in principle not be required to forgo their intellectual property'. Countries initially involved include France, Germany, the United Kingdom, Saudi Arabia, South Africa, Italy, Norway, Spain

---

[5] The Coalition for Epidemic Preparedness Innovations (CEPI) is a foundation endowed by donations from governments, philanthropic organisations and civil society organisations. It was established to fund independent research projects for the development of vaccines.

and Malaysia. **Many countries have not joined the Accelerator initiative** (European Commission, 2020c).

- Cyrus Poonawalla, the chief executive of the Serum Institute of India, the world's largest producer of vaccine doses, said the vaccine will have to first benefit the Indian nationals before it could be sent to other countries (Siddiqui, 2020). This is understandable considering the size of the Indian population. A bit problematic when you know that India has the highest capacity in the world for vaccine production.
- AstraZeneca reported that due to the $79 million investment from the UK, the first 30 million doses of the vaccine it is developing with the University of Oxford would be allocated to that country. Then, on 21 May 2020, the United States pledged as much as $1.2 billion to the company to obtain at least 300 million doses, with the first to be delivered as early as October 2020 (Marley, 2020).
- According to the map of COVID-19 temporary trade measures (11 June 2020):[6]
- Products affected by COVID-19 temporary export measures included personal protective equipment (e.g., masks, gloves), pharma products, hand sanitiser, food and certain other products.

    Export restrictions/bans (95 countries)
    Export liberalisations (2 countries)
    Export restrictions and liberalisations (3 countries)
    None (139 countries)

- The Donald Trump Administration reached a controversial agreement to take the entire global supply for the next 3 months of remdesivir (for which the result of published clinical trials do not show efficacy), one of the drugs being used in the treatment of COVID-19 (Yoo, 2020). The drug, produced by the US pharmaceutical company Gilead Sciences, is the first to be approved by US authorities for the treatment of the disease. According to the announcement by the Department of Health and Human Services, the agreement with Gilead guarantees 500,000 treatments, equivalent to 100% of July production, 90% of production in August and 90% of production in September (BBC News Mundo, 2020).

Never in the history of the pharmaceutical industry have such massive pledges of public funds for getting access to medicines or vaccines been seen. It is difficult to calculate and distinguish the sums channelled to the ACT Accelerator, CEPI, the WHO, Gavi and to the pharmaceutical companies themselves through the so-called 'Advance market commitments' (AMC). It is not very clear what the ownership status of the products resulting from these efforts will be. The costs and prices of the future vaccines are not clear either. Governments are buying and paying in advance for products that do not yet exist and whose safety and efficacy, if obtained, is uncertain.

Beyond the massive funding there is a need for a real global coordination capacity to ensure safety, efficacy of the products and equity in vaccine and treatment

[6] Map of COVID-19 temporary trade measures. https://www.macmap.org/covid19.

distributions according to well-defined priorities. Health workers and vulnerable people in all affected countries should be the first to receive the vaccine.

Ultimately, the race to develop and distribute a vaccine to prevent COVID-19 is overwhelmingly dominated by the private sector with a few large pharmaceutical companies playing a central role. How will we ensure that this 'commercial marathon' will end up with COVID-19 vaccines and related treatments that are safe and effective?

Massive public subsidies to the private sector, provided blindly, without clear conditions on products' characteristics, intellectual property protection, prices and distribution priorities, puts at risk the global solution that is needed. If the problem is planetary, the solution must be structured in a global way. Who will be the arbiter to avoid the 'vaccine nationalism'? (Sercovich, 2020). This should be the role of the WHO, but as the World Health Assembly of 18–19 May 2020 made it clear, industrialised countries are not willing to have WHO implement binding normative or governance instruments (Syam et al., 2020). Thus, the WHO recommends priorities for the distribution of the vaccines at national level: first health personnel, for example, then vulnerable people over 65, then people with other health problems. The key question is what can the WHO do to secure that those priorities are respected?

Several initiatives to address the COVID-19 health crisis have been launched or reformulated, as discussed next, by the WHO and other organisations. How can it be ensured that the global interests pursued by the WHO, the Global Fund, Gavi, CEPI, COVAX will not be overridden by national and commercial interests? According to *Le Monde*'s health specialist, Paul Benkimoun, 'The technological and financial battle being waged by the world's pharmaceutical companies to develop a vaccine is furious. It is a savage competition that suffers from a lack of collaboration and clear objectives' (Benkimoun et al., 2020, author translation). Future vaccines and treatments for COVID-19 are being considered as unprecedented commercial opportunities rather than a necessary tool to avoid suffering and deaths at a global scale in response to a humanitarian need.

One of the clear lessons of COVID-19 is, as noted, the interdependence of countries in the production of medicines and APIs. Interdependence that in cases of emergency can lead to a nationalistic response, which – aggravated by the inequalities between developed and developing countries – will inevitably lead to rich countries supplying themselves first. In this context, a central element of a well-articulated Global Preparedness for Health Emergencies would be the strengthening of local production of medicines and vaccines. It is local production that will be able to secure health sovereignty so that developing countries can ensure the availability of pharmaceuticals for prevention and treatment.

3.4 COVID-19 Global Vaccine Access Facility (COVAX Facility)

The WHO ACT Accelerator is a global collaboration to accelerate development, production, and equitable access to COVID-19 tests, treatments, and vaccines. Launched at the end of April 2020, it brings together Governments, scientists, businesses, civil society, and philanthropists and global health organisations.[7]

The ACT Accelerator launched a COVID-19 Global Vaccine Access Facility (COVAX Facility) in June 2020. The new facility will pool resources and share vaccine development risk. COVAX is co-led by Gavi,[8] the Coalition for Epidemic Preparedness Innovations (CEPI) and the WHO. Ninety-two low- and middle-income countries and economies will be able to access COVID-19 vaccines through Gavi COVAX Advance Market Commitment (AMC) (Gavi, 2020a).[9]

According to the WHO, 'demand guarantees for vaccine manufacturers will create access to substantial volumes of vaccines that will ultimately be safe and efficacious; better allocate capital; and support the manufacturing and procurement of sufficient volumes of vaccines to support equitable access globally. All countries will be invited to participate in the COVAX Facility. This investment opportunity of US$ 2 billion will provide vital seed funding to the Gavi Advance Market Commitment (AMC) for COVID-19 Vaccines (Gavi COVAX AMC) to support high-risk populations in low-income countries (LICs) and lower middle-income countries (LMICs), as part of the new COVAX Facility' (Gavi, 2020b).

As a result of the mechanism put in place, however, the COVAX Facility will enter into AMCs with the big pharmaceutical companies. This announcement has created a strong global reaction from various civil society organisations, particularly in developing countries, which are concerned about equitable access to future vaccines (Shashikant, 2020).

In June 2020, Gavi released a document titled 'The COVAX Facility: an insurance policy for COVID-19 vaccines' (Gavi, 2020b). Several aspects of this document are still unclear.

According to Third World Network (TWN), it is estimated that the proposed COVAX Facility will require funding of up to US$ 18.1 billion for the 2020/2021 vaccine supply (Shashikant, 2020). Of this total, US$ 11.3 billion is sought urgently to cover investments within the next 6 months, including US$ 2 billion in funding for advance market commitments to secure doses for LMICs. However, the justifications, including assumptions, for these estimates have not been provided. Conditions of how public funds will be used in advance market commitments are not known.

---

[7] See https://www.who.int/initiatives/act-accelerator.

[8] Gavi's beneficiary countries have only included the poorest, those with a Gross National Income (GNI) per capita of less than or equal to US$ 1580. See https://www.gavi.org/types-support/sustainability/eligibility.

[9] See https://www.gavi.org/news/media-room/92-low-middle-income-economies-eligible-access-covid-19-vaccines-gavi-covax-amc.

The Gavi COVAX AMC will produce a supply of vaccines for LICs and LMICs. It is unclear what terms and conditions will be attached to the financial instruments for developing countries. In short, the COVAX facility prioritises the needs of self-financing countries that participate in its scheme. On pricing, the proposal states 'flat pricing strategy...will be encouraged', but firms are free to set their own prices (Gavi, 2020a).

As noted by one commentator, 'Demand for a particular vaccine (albeit with unproven effectiveness) through various competing advance purchase agreements (the COVAX facility, the European Union and United States agreements), each presumably trying to outbid one another, only serves to benefit the pharmaceutical industry's profiteering through high prices' (Shashikant, 2020).

The proposal states that the 'Facility has access to doses of vaccine candidates through agreements that provide manufacturer-specific contingent volume guarantees to procure vaccines that meet WHO Target Product Profile to de-risk and incentivise timely investment in expansion of manufacturing capacity' (Gavi, 2020a).

A recent Médecins Sans Frontières (MSF) paper points out that Gavi is a Swiss-based foundation with a mandate to finance vaccines for the world's poorest countries – currently 58 eligible countries (of an original 73 eligible countries) (MSF, 2020a). However, it questions Gavi's role in hosting a global 'facility' for COVID-19 vaccines, which 'is beyond the organization's mandate and expertise', stressing that 'Gavi has no experience working with most MICs nor any high-income countries (HICs) on procuring for the countries' vaccine needs' and 'does not have experience negotiating with pharmaceutical companies on behalf of these countries'.

On 23 June 2020, 45 civil society organisations sent a letter to the board members of Gavi highlighting concerns about the fact that 'pharmaceutical companies are allowed to retain and pursue rights to vaccines under development, resulting in vaccines that are proprietary and under the monopoly control of individual companies. Since there has been no change in how intellectual property is handled during the pandemic, pharmaceutical companies are able to monopolize future COVID-19 vaccines and decide who does and does not get access' (MSF, 2020b). The letter points out that more than US$ 4.5 billion of public and philanthropic funding has already been given to companies for COVID-19 vaccine research and development (R&D), and 'Gavi is now designing a fund to award further money to pharmaceutical corporations'. It further notes that 'The public and philanthropic funding already awarded should result in the delivery of effective vaccines that are designated as global public goods: sold at cost and free from monopoly control', and suggests a number of criteria to finance, price and allocate vaccines (MSF, 2020b).

These concerns justifiably focus on the equitable access to the vaccines to be developed. There is, however, a need to address other facets of the current situation in terms of the participation of developing countries not just as mere recipients of vaccines made abroad but as partners in their production. In fact, part of the response to the current supply crisis should be the creation or strengthening of vaccine production capacity in developing countries. Why not to think about a modality of AMCs with developing countries' producers that have the capacity to manufacture the new vaccines? Why not to support the technological upgrading of plants in those countries to expand the global capacity to respond to this and future vaccination needs? Why not put in place a programme for building additional manufacturing capacity in developing countries in this field to overcome the current oligopolistic market for vaccines? These actions should be based on the understanding that a vaccine in time of a pandemic should be in the public domain and considered a global public good (MSF, 2020a).

In summary, the current response to the development and production of vaccines to address COVID-19 raises many questions and concerns. More attention should be given to the potential role of developing countries in the production (and not only consumption) of such vaccines and on the policy measures that would need to be adopted (as developed countries are doing now) to ensure greater autonomy in their supply as well as to increase those countries' participation in the global production.

3.5   Global Preparedness Monitoring Board

The Global Preparedness Monitoring Board (GPMB) is an independent monitoring and advocacy body. It urges political action to prepare for and mitigate the effects of global health emergencies (WHO, 2019). Co-convened in May 2018 by the World Bank Group and the World Health Organization, the Board builds on the work of the Global Health Crises Task Force and Panel, created by the United Nations Secretary-General in the wake of the 2014–2016 Ebola epidemic. The Board works independently of all parties, including its co-conveners, to provide the frankest assessments and recommendations possible. The 15-member Board is made up of political leaders, heads of agencies, and experts, led jointly by Dr. Gro Harlem Brundtland, formerly Prime Minister of Norway and Director-General of the World Health Organization and Mr. Elhadj As Sy, Secretary General of the International Federation of Red Cross and Red Crescent Societies. The goals of the Board are to assess the world's ability to protect itself from health emergencies, identify critical gaps to preparedness across multiple perspectives; advocate for preparedness activities with national and international leaders and decision-makers.

The Preparedness Monitoring Board of the WHO and the World Bank reviewed recommendations from previous high-level panels and commissions following the 2009 H1N1 influenza pandemic and the 2014–2016 Ebola outbreak, along with its own commissioned reports and other data.

The recommendations in this report relate to the following seven points, *one of which (point 4) speaks on ensuring adequate investment in developing innovative vaccines and therapeutics and surge manufacturing capacity.* This report, published in 2019, was supposed to capture all the experiences and lessons about pandemic preparedness, but it does not mention anything about possible 'vaccine nationalism'. Nor does it anticipate what COVID-19 has highlighted, such as the need to strengthen production capacity in developing countries. Here are the seven points of the report (WHO, 2019):

1. **Heads of Government must commit and invest.**

   Heads of Government in every country must commit to preparedness by implementing their binding obligations under the International Health Regulations (IHR) (2005).

2. **Countries and regional organisations must lead by example.**

   G7, G20 and G77 Member States and regional intergovernmental organisations must follow through on their political and funding commitments for preparedness.

3. **All countries must build strong systems.**

   Heads of Government must appoint a national high-level coordinator with authority to maintain effective preparedness.

4. **Countries, donors and multilateral institutions must be prepared for the worst.**

   A rapidly spreading pandemic due to a lethal respiratory pathogen (whether naturally emergent or accidentally or deliberately released) poses additional preparedness requirements. Donors and multilateral institutions must ensure adequate investment in developing innovative vaccines and therapeutics, surge manufacturing capacity, broad-spectrum antivirals and appropriate non-pharmaceutical interventions. All countries must develop a system for immediately sharing genome sequences of any new pathogen for public health purposes along with the means to share limited medical countermeasures across countries.

5. **Financing institutions must link preparedness with economic risk planning.**

   To mitigate the severe economic impacts of a global pandemic, the International Monetary Fund (IMF) and the World Bank must urgently renew their efforts to integrate preparedness into economic risk and institutional assessments.

6. **Development assistance funders must create incentives and increase funding for preparedness.**

   Donors, international financing institutions, global funds and philanthropies must increase funding for the poorest and most vulnerable countries through development assistance for health and greater/earlier access to the United Nations Central Emergency Response Fund to close financing gaps for their national actions plans for health security.

7. **The United Nations must strengthen coordination mechanisms.**

   The WHO should introduce an approach to mobilise the wider national, regional and international community at earlier stages of an outbreak, before a declaration of an IHR (2005) Public Health Emergency of International Concern (PHEIC).

These recommendations for pandemic preparedness due to a lethal respiratory pathogen have been re-stated, annually in the WHO documents and resolutions since the 2009 H1N1 pandemic. If they had been taken seriously, there would have been no shortage of masks and breathing apparatus at the beginning of 2020, and the capacity to produce vaccines would have been increased. We currently do not have the vaccine, but we also know that if the vaccine arrives tomorrow, we do not have enough production capacity installed.

3.6   A COVID-19 Technology Sharing Platform: A Recent UN Initiative[10]

In May 2020, the United Nations Technology Bank, together with the UNDP, UNCTAD, and the WHO launched the Tech Access Partnership (TAP) as part of a coordinated approach to strengthen developing countries' responses to COVID-19 and increase access to lifesaving health technologies (United Nations, 2021).

TAP aims to address critical shortages of essential health technologies and equipment by connecting manufacturers with critical expertise and emerging manufacturers in developing countries to share the information, technical expertise and resources necessary to scale up production of these tools. The Partnership will also support countries to develop affordable technologies and equipment that meet quality and safety standards.

'Now, more than ever, the global community needs to unite to save lives and secure sustainable futures. Inequalities are exacerbating the technology and digital divide when it comes to opportunities for youth, creating a divide that threatens to leave them behind', says Amina J. Mohammed, Deputy Secretary-General of the UN. 'Increasing access to necessary technologies through partnerships is a crucial component of the United Nation's COVID-19 health, humanitarian and socio-economic response' (UNDP, 2020).

TAP will be led by the UN Technology Bank for Least Developed Countries, established in 2016 to assist governments with the development and adaptation of new technologies. The initiative, which is open to all developing countries, will also be supported by its core partners, UNDP, UNCTAD and the WHO.

The key functions of TAP will include:

- Product information – a digital warehouse of manufacturing and design specifications, technical knowledge and information required to increase capacity.
- Technical Guidance – a technical support line to help manufacturers.
- Partnerships – a platform to match companies based on expertise, needs and capacity (UNDP, 2020).

TAP aims *to supports developing countries to scale up local production of critical health technologies needed to combat COVID-19*, including personal protective equipment, diagnostics and medical devices such as ventilators.

This UN initiative seems to confirm that COVID-19 requires rethinking of local production in developing countries.

---

[10] See A COVID-19 Technology Sharing Platform, https://www.un.org/technologybank/news/un-agencies-launch-tech-access-partnership-joint-effort-scale-local-production-life-saving.

3.7 Concluding Remarks

During the 73rd World Health Assembly (WHA) in May 2020, the United Nations Secretary-General and several Heads of State made important declarations of principle. These declarations stressed that all possible treatments, present and future, including vaccine(s) related to the COVID-19 pandemic, should be considered as global public goods and should be available, to all, at the same time and in sufficient quantities. These statements are not viable when they clash with the reality of how the global pharmaceutical market is organised and with growing nationalistic trends on the production and distribution of vaccines to address the pandemic.

The 73rd WHA was a little paradoxical, full of solemn declarations and a few substantial financial pledges, without precedent, while at the same time an unambitious resolution, 'COVID-19 response', was approved. The resolution was far from containing clear instruments to put into practice the intentions expressed in the solemn declarations.

The COVID-19 pandemic has shown renewed efforts by developed countries to ensure autonomy in the manufacturing of pharmaceuticals and has given rise to nationalistic approaches. At the same time, it is clear that even if one or more vaccines against COVID-19 are successfully developed, there is no sufficient global manufacturing capacity to produce the billions of doses needed to protect the world population. In this context, it seems urgent to reopen the discussion about the local pharmaceutical production and how developing countries can expand their capacity to participate in the global market for APIs and pharmaceuticals, including biologicals. A portion of the public funds in the form of AMCs should go to developing countries that have the technological capacity to produce vaccines.

References

Bayer, L., Deutsch, J., Hanke Vela, J., & Tamma, P. (2020). EU moves to limit exports of medical equipment outside the bloc. *Politico*, 15 March 2020. https://www.politico.eu/article/coronavirus-eu-limit-exports-medical-equipment/.

BBC News Mundo. (2020). Remdesivir: la polémica compra de EE. UU. de casi toda la existencia mundial del prometedor fármaco para combatir el covid-19. 1 July 2020. https://www.bbc.com/mundo/noticias-internacional-53254231.

Benkimoun, P., Delacroix, G., Hecketsweiler, C., & Herzberg, N. (2020). A la recherche du vaccin contre le Covid-19: la course acharnée entre les laboratoires et les Etats. *Le Monde*, 24 June 2020. https://www.lemonde.fr/planete/article/2020/06/24/coronavirus-la-guerre-sans-merci-des-laboratoires-pour-un-vaccin_6043964_3244.html.

Bezat, J.-M. (2020). Le gouvernement amorce une politique de relocalisation des médicaments. *Le Monde,* 19 June 2020. https://www.lemonde.fr/economie/article/2020/06/18/le-gouvernement-amorce-une-politique-de-relocalisation-des-medicaments_6043337_3234.html.

Brennan, Z. (2020). The next unprecedented vaccine hurdle: Making hundreds of millions of doses. *Politico*, 13 August 2020. https://money.yahoo.com/next-unprecedented-vaccine-hurdle-making-235507169.html.

Capital. (2020). Emmanuel Macron veut relocaliser en France des productions critiques de médi-caments. 17 June 2020. https://www.capital.fr/economie-politique/emmanuel-macron-veut-relocaliser-en-france-des-productions-critiques-de-medicaments-1372801.

Correa, C. (2020). Lessons from COVID-19: Pharmaceutical production as a strategic goal. SOUTHVIEWS No. 202. South Centre, July 2020. https://www.southcentre.int/wp-content/uploads/2020/07/SouthViews-Correa.pdf.

European Commission. (2020a). Press release, Coronavirus Global Response: €7.4 billion raised for universal access to vaccines Brussels. 4 May 2020. https://ec.europa.eu/commission/presscorner/detail/en/ip_20_797.

European Commission. (2020b). Press release, Pharmaceutical strategy: European Commission launches open public consultation. 16 June 2020. https://ec.europa.eu/commission/presscorner/detail/en/ip_20_1065.

European Commission. (2020c). Questions and answers: the coronavirus global response. Brussels, 28 May 2020 https://ec.europa.eu/commission/presscorner/detail/en/qanda_20_958.

Gavi. (2020a). 92 low- and middle-income economies eligible to get access to COVID-19 vac-cines through Gavi COVAX AMC. Geneva, 31 July 2021. https://www.gavi.org/news/media-room/92-low-middle-income-economies-eligible-access-covid-19-vaccines-gavi-covax-amc.

Gavi. (2020b). The Gavi COVAX AMC: An investment opportunity. June 2020. https://www.Gavi.org/sites/default/files/2020-06/Gavi-COVAX-AMC-IO.pdf.

Gavi. (2020c). Global vaccine summit on 4 June 2020, "Chair's Summary". https://www.Gavi.org/sites/default/files/2020-06/4-June-2020-Global-Vaccine-Summit-Gavi-3rd-Replenishment-Chairs-Summary.pdf.

GTAI Germany Trade & Invest. (2020). Covid-19 fuels debate over API production loca-tions. 14 April 2020. https://www.gtai.de/gtai-en/invest/industries/life-sciences/covid-19-fuels-debate-over-api-production-locations-239724.

Kaplan, W., & Laing, R. (2005). Local production of pharmaceuticals: industrial policy and access to medicines – An overview of key concepts, issues and opportunities for future research (English). Health, Nutrition and Population (HNP) discussion paper. Washington, DC: World Bank. http://documents.worldbank.org/curated/en/551391468330300283/Local-production-of-pharmaceuticals-industrial-policy-and-access to medicines-an-overview-of-key-concepts-issues-and-opportunities-for-future-research.

Kayali, L. (2020). Russia produces first batch of its coronavirus vaccine. *Politico*, 15 August 2020. https://www.politico.eu/article/russia-begins-production-of-coronavirus-vaccine/.

Lavarello, P., Gutman, G., & Sztulwark (Eds.). (2018). *EXPLORANDO EL CAMINO DE LA IMITACIÓN CREATIVA: La industria biofarmacéutica Argentina en los 2000*. Libro, CEUR-CONICET.

Le Monde avec AFP. (2020). Sanofi et un vaccin contre le Covid-19 en priorité pour les Etats-Unis: une polémique vite devenue politique en France. *Le Monde*, 15 May 2020. https://www.lemonde.fr/sante/article/2020/05/14/vaccin-contre-le-covid-19-inacceptable-que-sanofi-serve-en-premier-les-etats-unis_6039621_1651302.html.

Lopez Girondo, A. (2020). El COVID 19 le hizo ver a la UE su dependen-cia en la industria farmacéutica. May 2020. https://www.tiempoar.com.ar/nota/el-covid-19-le-hizo-ver-a-la-ue-su-dependencia-en-la-industria-farmaceutica.

Marley, S. (2020). AstraZeneca aims for 30 million U.K. vaccine doses by September. *Bloomberg*, 17 May 2020. https://www.bloomberg.com/news/articles/2020-05-17/astrazeneca-aims-for-30-million-u-k-vaccine-doses-by-september.

Milne, R., & Vrow, D. (2020). Why vaccine 'nationalism' could slow coronavirus fight. *Financial Times*, 14 May 2020. https://www.ft.com/content/6d542894-6483-446c-87b0-96c65e89bb2c.

Moore, J. (2020). Vaccine nationalism is unfair and unwise. By putting themselves at the front of the line for COVID-19 shots, the US and other countries might make poorer nations wait years for theirs. It does not have to be this way, Globe Ideas. https://www.bostonglobe.com/2020/08/29/opinion/vaccine-nationalism-is-unfair-unwise/.

MSF. (2020a). COVID-19 vaccine global access (COVAX) facility: Key considerations for Gavi's Global financing mechanism. June 2020. https://msfaccess.org/sites/default/files/2020-06/MSF-AC_COVID-19_Gavi-COVAXFacility_briefing-document.pdf.

MSF. (2020b). Open letter to Gavi Board Members: Inclusion of civil society in COVAX Facility and COVAX AMC governance is essential. 29 July 2020. https://msfaccess.org/sites/default/files/2020-06/Vax_LetterToGaviBoard_22June2020-final__0.pdf.

Santos Rutschman, A. (2020). How 'vaccine nationalism' could block vulnerable populations' access to COVID-19 vaccines. 17 June 2020. https://theconversation.com/how-vaccine-nationalism-could-block-vulnerable-populations-access-to-covid-19-vaccines-140689.

Sercovich, F. (2020). Coronavirus pandemic: the vaccine as exit strategy. A global hurdle race against time with a split jury. SOUTHVIEWS No. 203, 24 July 2020. https://www.southcentre.int/southviews-no-203-24-july-2020/.

Shashikant, S. (2020). COVID-19: global concern that Gavi's vaccine initiative promotes inequitable access. Third World Network, 29 June 2020.

Siddiqui, Z. (2020). India's Serum Institute to make millions of potential coronavirus vaccine doses. *Nasdaq*, 29 April 2020. https://www.nasdaq.com/articles/indias-serum-institute-to-make-millions-of-potential-coronavirus-vaccine-doses-2020-04-29.

Stork, M., & Wanandi, S. (1980). *Guidelines and recommendations for the establishment of a low cost pharmaceutical plant* (p. 72). UNCTAD.

Syam, N. (2020). EU Parliament adopts resolution on public health strategy post-COVID-19 based on use of TRIPS flexibilities to ensure access to health technologies, SOUTHVIEWS No. 329. South Centre, 12 August 2020. https://us5.campaign-archive.com/?u=fa9cf38799136b5660f36 7ba6&id=dc238cfbb4.

Syam, N., Alas, M., & Ido, V. (2020). The 73rd World Health Assembly and Resolution on COVID-19: Quest of global solidarity for equitable access to health products, Policy Brief No. 78. South Centre, May 2020. https://www.southcentre.int/policy-brief-78-may-2020/.

Thanh Le, T., Kramer, J. P., Chen, R., & Mayhew, S. (2020). Evolution of the COVID-19 development landscape. *Nature Reviews Drug Discovery*, 7 September 2020. https://www.nature.com/articles/d41573-020-00151-8.

UK Government. (2020). Press release, UK Government secures new COVID-19 vaccines and backs global clinical, trial. 14 August 2020. https://www.gov.uk/government/news/uk-government-secures-new-covid-19-vaccines-and-backs-global-clinical-trial.

UNCTAD. (2011). Local production of pharmaceuticals and related technology transfer in developing countries. A series of case studies by the UNCTAD Secretariat. https://unctad.org/en/PublicationsLibrary/diaepcb2011d7_en.pdf.

UNCTAD. (2017). Tool box for policy coherence in access to medicines and local pharmaceutical production. https://unctad.org/en/PublicationsLibrary/diaepcb2017d2_en.pdf.

UNDP. (2020). UN agencies launch Tech Access Partnership in joint effort to scale up local production of life-saving health technologies for COVID-19. New York, 12 May 2020. https://www.undp.org/content/undp/en/home/news-centre/news/2020/UN_agencies_launch_Tech_Access_Partnership_in_joint_effort_to_scale_up_local_production_of_life-saving_health_technologies_for_COVID-19.html.

UNIDO. (1980). La croissance de l'industrie pharmaceutique dans les pays en développement: Problèmes et perspectives. New York.

UNIDO. (2020). Pharmaceutical production in developing countries. Vienna. https://www.unido.org/our-focus-advancing-economic-competitiveness-investing-technology-and-innovation-competitiveness-business-environment-and-upgrading/pharmaceutical-production-developing-countries.

United Nations. (2021). UN agencies launch Tech Access Partnership in joint effort to scale up local production of life-saving health technologies for COVID-19. 3 June 2021. https://www.un.org/technologybank/news/un-agencies-launch-tech-access-partnership-joint-effort-scale-local-production-life-saving.

Velásquez, G. (1986). *Salud, medicamentos y Tercer Mundo*. Madrid, IEPALA.

Velásquez, G. (2015). Managing an A(H1N1) pandemic: Public health or healthy business. In A. Karan & G. Sodhi (Eds.), *Protecting the health of the poor*. Zed Books.

Velásquez, G. (2020). Rethinking R&D for pharmaceuticals products after the novel coronavirus COVID-19 shock, Policy Brief No. 75. South Centre, April 2020. https://www.southcentre.int/policy-brief-75-april-2020/.

Weintraub, R., Bitton A., & Rosenberg M. L. (2020). The danger of vaccine nationalism. *Harvard Business Law Review*, June 2020. https://hbr.org/2020/05/the-danger-of-vaccine-nationalism.

WHO. (2007). Pandemic influenza vaccine supplies = Production de vaccins contre la grippe pandémique. *Weekly Epidemiological Record = Relevé épidémiologique hebdomadaire, 82*(45), 399–400. https://apps.who.int/iris/handle/10665/241035

WHO. (2008). WHA 61.21 The global strategy and plan of action on public health, innovation and intellectual property. https://www.who.int/phi/implementation/phi_globstat_action/en/.

WHO. (2011a). Local production and access to medicines in Low- and middle-income countries A literature review and critical analysis. Geneva. https://www.who.int/phi/publications/Local_Production_Literature_Review.pdf?ua=1.

WHO. (2011b). Preparación para una gripe pandémica: Marco para el intercambio de virus gripales y el acceso a las vacunas y otros beneficios. https://apps.who.int/iris/bitstream/handle/10665/44867/9789243503080_spa.pdf;jsessionid=83D46BFFE59F6AB5C5F423EF6A0D1474?sequence=1.

WHO. (2017). Report of the interagency consultation on local production of essential medicines and health products. 25 April 2017. https://apps.who.int/iris/bitstream/handle/10665/255847/WHO-EMP-2017.02-eng.pdf?sequence=1.

WHO. (2019). Global Preparedness Monitoring Board, *A WORLD AT RISK, Annual report on Global preparedness for health emergencies*. : WHO. https://apps.who.int/gpmb/assets/annual_report/GPMB_Annual_Report_English.pdf.

Wintour, P. (2020). World leaders pledge €7.4bn to research Covid-19 vaccine. *The Guardian*, 4 May 2020. https://www.theguardian.com/world/2020/may/04/world-leaders-pledge-74bn-euros-to-research-covid-19-vaccine.

Yoo, J.-H. (2020). Uncertainty about the efficacy of Remdesivir on COVID 19. *Journal of Korean Medical Science*, 10 June 2020. https://doi.org/10.3346/jkms.2020.35.e221.

Chapter 4
Rethinking R&D for Pharmaceutical Products After the Novel Coronavirus COVID-19 Shock

4.1 Introduction

The unprecedented global health crisis caused by the COVID-19 pandemic, during the first quarter of 2020, brings back with particular urgency the discussion about the research and development (R&D) model for pharmaceuticals and other technologies necessary to respond to the health problems of both developed and developing countries.

This chapter argues that the current R&D model for pharmaceutical products is fragmented, inefficient, expensive, and full of overlaps and waste of resources, and that it will not be able to provide the global solution that the COVID-19 crisis requires. A new R&D model based on health rather than commercial interests–generally supported on patents and other intellectual property rights– can be designed and implemented under the auspices of the World Health Organization (WHO) based on Article 19 of the WHO Constitution.

Section 4.1 of this chapter refers to the background of the debate on the R&D model for pharmaceutical products and other health technologies. Section 4.2 addresses some of the problems of the current R&D model. Section 4.3 briefly summarises what could be the objectives and principles of a binding convention on R&D. Section 4.4 argues that there is a need to move fast and discusses some recent initiatives. Finally, what would be the way forward is briefly considered.

We were warned... Was the current health crisis foreseeable? Was there any indication that a phenomenon of this nature could happen?

In May 2011, a WHO document on pandemic influenza preparedness alerted countries about the continuing risk of an influenza pandemic with potentially

---

This chapter is largely taken from: Velásquez, G. (2020 April). *Rethinking R&D for Pharmaceutical Products After the Novel Coronavirus COVID-19 Shock*. South Centre Policy Brief No. 75. https://www.southcentre.int/wp-content/uploads/2020/04/PB-75-Rethinking-RD-after-COVID-19-Shock-REV.pdf.

© SC: South Centre 2022

G. Velásquez, *Vaccines, Medicines and COVID-19*, SpringerBriefs in Public Health, https://doi.org/10.1007/978-3-030-89125-1_4

devastating health, economic and social consequences, particularly for developing countries, which have a higher disease burden and are more vulnerable (WHO, 2011). The 2019 Annual Report on Global Preparedness for Health Emergencies, prepared by the World Bank's Global Preparedness Monitoring Board, referred to 'a very real threat of a rapidly moving, highly lethal pandemic of a respiratory pathogen killing 50 to 80 million people and wiping out nearly 5% of the world's economy' (WHO, 2019). This indicates that experts[1] have been anticipating the risk of a pandemic such as the one we are experiencing now (Carrington, 2020). Why were these warnings not heeded?

Noam Chomsky recently said about the outbreak of COVID-19: 'The neoliberal assault has left hospitals unprepared. One example among many: hospital beds have been suppressed in the name of efficiency (…). This crisis is the umpteenth example of market failure, just as the threat of environmental catastrophe is. The governments and the pharmaceutical multinationals companies have known for years that there is a high probability of a serious pandemic, but since it is not good for the profits to prepare for it, nothing has been done' (Nicoli, 2020).

Recent data on the Italian situation confirms well with Chomsky's statement. In Italy, one of the most affected countries by the coronavirus crisis, 'in less than ten years, from 2010 to 2016, 70,000 hospital beds disappeared, 175 hospital units were closed, and local autonomous health offices were reduced from 642 in the 1980s to only 101 in 2017. All of this is for the benefit of the private health and insurance industries, which offer no protection against pandemics' (Nicoli, 2020).

If the imminent arrival of 'an influenza pandemic with potentially devastating health, economic and social consequences' was already mentioned in WHO documents since 2011, why 10 years after the arrival of the current crisis, there was no complete mapping of what the R&D situation was in terms of vaccines and treatments? The 'Solidarity' clinical trial for COVID-19 treatments was launched by the WHO Director General on 18 March 2020 almost 3 months after the start of the problem, but too late to provide a fast response to the already devastating effects of the coronavirus (WHO, 2020).

And how the global production and distribution of the vaccine will be organised when it arrives? Will the detainment of products in transit, trade restrictions, the enforcement of intellectual property rights be allowed to prevail over global public health interests? Who is going to make the rules to ensure that the vaccine reaches everybody, in all places at the same time? Who is going to enforce them? Who will protect the world's public interest?

It is time to develop multilateral rules and empower the WHO so that it can exercise a real global coordination on health matters: COVID-19 has unveiled the shortcomings of global governance in public health. States must work together and in a coordinated manner to face the new threats and secure fair and adequate access to medicines for all (Barbancey, 2020).

---

[1] A 2007 study of the 2002–03 SARS outbreak concluded that the presence of a large reservoir of Sars-CoV-like viruses in horseshoe bats, together with the culture of eating exotic mammals, was 'a timebomb'.

4.2   Background of the Debate on the R&D Model

In May 2012, the WHO Member States meeting at the World Health Assembly in Geneva, adopted resolution WHA 65.22 endorsing the recommendations of the Consultative Expert Working Group on Research and Development: Financing and Coordination (CEWG). For many of the World Health Assembly (WHA) participants and observers those recommendations provided a roadmap for a first step towards a change in the current pharmaceutical R&D model for pharmaceutical products. Particularly on the premise that the market cannot be the only driver of R&D, the CEWG recommended the negotiation of an international convention in which all countries would commit to promote R&D: 'formal intergovernmental negotiation should begin for a binding global instrument for R&D and innovation for health' (World Health Assembly 65, 2012).

In fact, to ensure a sustainable long-term R&D and subsequent affordable access to pharmaceuticals in developing as well as developed countries, rather than adopt voluntary guidelines or recommendations, the WHO should use its capacity to legislate. Negotiating and adopting an R&D convention is one the paths to follow. If it were in place now, there would have been a much greater capacity and better tools to address the current health crisis.

It is time to develop and better use international health law to effectively address the global health problems. Under Article 19 of the WHO Constitution:

> The Health Assembly shall have authority to adopt conventions or agreements with respect to any matter within the competence of the Organization. A two-thirds vote of the Health Assembly shall be required for the adoption of such conventions or agreements, which shall come into force for each Member when accepted by it in accordance with its constitutional processes (WHO, 2006a).

The protection of health in times of global health crisis risks reflects a pressing social need that should now be translated into the vocabulary of international law. While enormous challenges lie ahead, especially in terms of the use and strengthening the existing instruments, as noted by one commentator, '[t]here is an urgent need for counterbalancing interests such as international trade, global commerce and the welfare interests of the protection of the health of both individuals and populations worldwide' (Toebes, 2015).

Article 19 of the WHO Constitution is the best example of existing international health law, which has already been successfully tested in the case of the WHO Framework Convention on Tobacco Control (FCTC). Tobacco is the first killer in the world. The FCTC is the most efficient global instrument negotiated in the WHO: it has become the 'vaccine' against cancer and cardiovascular diseases (Velásquez & Seuba, 2011, p. 8).

Despite the regulatory powers conferred by its constitution under Article 19, 'WHO has paid but scarce attention to law – especially the *hard law* – as a tool to protect and promote health. On the contrary, the Organization has shown itself to be more in favour of seeking a political agreement and has excused itself in its medicosanitary profile in order to take on more of a health care than a legal role' (Seuba,

2010). The FCTC is the only case in which said article has been used in the history of the WHO.

In the present international context of the COVID-19 pandemic, the WHO may recover its leadership through the use of Article 19 of its constitution by negotiating and adopting global treaties and conventions to help Members States to realise the right to access to health, including in situations of global emergencies, and to achieve the Universal Health Coverage (UHC) (Seuba, 2010, p. 58).

The directives and technical recommendations of the WHO, which are relevant and appropriate in most cases, often are not heeded or followed because they are only recommendations of a voluntary nature. The countries of the European Union, for instance, were unable to agree on the common strategy recommended by the WHO against the coronavirus pandemic. In cases of global health crises, it is essential that necessary measures can be made binding and enforceable. Pandemics have no borders. While the WHO could not take compulsory measures, many countries did, and it would have been more consistent if solid WHO guidelines had been mandatory via Article 19 of the WHO Constitution, or the International Health Regulations.

The aim of an international convention would primarily be to set up an international public fund for pharmaceutical R&D. To ensure sustainability of the fund, the convention would need to provide for a mandatory contribution by signatory countries according to their level of economic development. In return, the products and results financed by this fund would be considered as public goods benefiting all these countries. Hence, the idea is not just to require another financial contribution but rather to put in place an innovative mechanism that better focuses on patients' interests than under the current R&D model. Moreover, the costs of R&D activities financed by this public fund would have to be transparent to guarantee a more efficient and less costly innovation system that meets the real sanitary needs of countries of both the Global North and the Global South. Should such mechanism be in place, it would have facilitated to provide a global financial support for the development of products for prevention and treatment of COVID-19 by those capable of undertaking the needed R&D (Lurie et al., 2020). If an international convention, as proposed, with its financial mechanism, would have been in place, the task would have been easier and accomplished faster.

A binding international convention, negotiated under the auspices of the WHO, could thus serve to sustainably finance R&D on useful and safe pharmaceuticals to respond to public health needs, at prices affordable to patients and health systems. Moreover, the adoption of a convention of this nature under Article 19 of the WHO Constitution, could be the prelude to a deeper reflection on world health governance (Smolar, 2020).[2]

---

[2] In an article in the newspaper 'Le Monde' on 31 March 2020 by Piotr Smolar, it is argued that one of the sectors that will have to be rethought is the R&D model for health products and the role of the pharmaceutical industry. https://www.lemonde.fr/international/article/2020/03/31/coronavirus-comment-la-diplomatie-francaise-pense-le-jour-d-apres_6034979_3210.html.

The negotiation and adoption of an international treaty on health R&D was one of the key elements for the implementation of the *Global Strategy* on Public Health, Innovation and Intellectual Property (*GSPOA*). Indeed, if successful, this could have been the most important achievement of the GSPOA (Velásquez, 2019).

4.3 Problems of the R&D Model for Pharmaceutical Products[3]

The current R&D model for pharmaceutical products is based on the following scheme: *Research* (private or public) – *patents* (legal monopoly) – *high prices – restricted access.*[4] This model presents several problems that eventually led to a disarticulation between innovation and access. These problems include: (1) Lack of transparency of R&D costs; (2) net decrease of pharmacological innovation in the last years.; (3) high prices restricting access.; (4) fragmentation and lack of coordination; and (5) waste and overlap.

4.3.1 Lack of Transparency of R&D Costs

The cost estimated by a study of Boston Tufts Centre, for the development of a new molecule was of US$ 2.5 billion (Tufts Centre, 2014). This is the figure used since then by the 'originator' pharmaceutical industries to argue about the high costs they incur and the need for high prices of medicines to recover them. However, in a study carried out by the London School of Economics, the authors found that the average cost to develop a new product was only US$ 43.4 million (Light & Warburton, 2011). The non-profit foundation Drugs for Neglected Diseases initiative (DNDi) reported in 2019 that the cost for research and development of a sleeping sickness drug was US$ 55 million (DNDi, 2019).

As long as there is no clarity on the real cost of R&D, the problem of prices – and therefore of access to medicines – will continue to go unsolved. The massive difference between the estimates of US$ 55 million or US$ 2.5 billion per molecule clearly indicates that the resulting prices of new medicines, if reasonably based on real R&D costs, would be significantly different.

---

[3]This section is partially based on: Velásquez, G., & Seuba, X. (2011 December). *Rethinking Global Health: A Binding Convention for R&D for Pharmaceutical Products*. South Centre Research Paper 42. https://www.southcentre.int/wp-content/uploads/2013/04/RP42_Rethinking-global-health_EN.pdf

[4]All members of the World Trade Organization (WTO) are bound to grant patents for pharmaceuticals.

4.3.2   Pharmaceutical Innovation Has Significantly Decreased

The number of new molecules approved for therapeutic use has declined in the last two decades despite the advancement of science and technology and the availability of financial resources to undertake R&D for the diseases prevailing in developed countries. In addition, the therapeutic value of most of the new medicines has also gone down. According to data published by the French magazine *Prescrire,* for instance, the average of the number of drugs, representing 'a major therapeutic advance' introduced on the French market in 10 years (2007–2017) was 4.7 products per year. But these numbers decreased significantly, from 14 products in 2007 to only one product in 2017 (Prescrire, 2017). 'The number of new drugs approved per billion US dollars spent on R&D has halved roughly every 9 years since 1950, falling around 80-fold in inflation-adjusted terms' (Connell et al., 2012).

In the area of therapeutics for cardiovascular diseases (CVD), for instance, Gail A. Van Norman describes adverse trends towards declining innovation and rising costs of drug development over the last several decades. 'Thirty-three percent fewer CVD therapeutics were approved between 2000 and 2009 compared to the previous decade, and the number of CVD drugs starting all clinical trial stages declined in both absolute and relative numbers between 1990 and 2012. In the last 5 years, drugs to treat CVD disease comprised just 6 per cent of all new drug launches' (Van Norman, 2017).

According to a recent study by *STAT Reports*, major pharmaceutical manufacturers are not the originators of the majority of the medicines they sell. In fact, it appears that they had already scaled back their expenditure on research for new drugs (Jung et al., 2019).

4.3.3   High Prices Restrict Access

In 2014, the American firm Gilead Sciences introduced the hepatitis C drug sofosbuvir (brand name SOVALDI®) at the eye-watering price in the USA of US$ 84,000 for a 12-week treatment. In 2015 the American firm Vertex introduced Orkambi®, a medicine used to treat cystic fibrosis in patients ages 2 years and older, at the price of US$ 272,000 per patient per year. A study in the US on 71 anti-cancer medicines approved between 2002 and 2014 by the US Food and Drug Administration (FDA) found that many of them cost more than US$ 100,000 per treatment per year (Tibau et al., 2016). In 2018 Novartis introduced the CAR-T leukaemia treatment Kymriah® at US$ 350,000. On 27 May 2019 the US FDA gave marketing authorisation for 'Zolgensma®' a gene therapy, also from Novartis. The price of the drug, administered in a single dose, is US$ 2125 million, making it the most expensive drug in the history of the pharmaceutical industry (Velásquez, 2019).

This escalation of prices over the last 5 years, especially for products of biological origin, has been recently justified by the industry on the argument that prices should be set based on the 'value' of the product for the patient rather than on the cost of R&D, as was previously the case. Neither governments nor the WHO have challenged this new concept so far, which is not practiced in any other manufacturing sector, except perhaps in luxury industries.

Lack of transparency on the costs of R&D, a diminishing rate of pharmaceutical innovation in recent years and high prices, in conjunction, demonstrate that a structural problem exists in the current R&D model for pharmaceutical products. Several documents discussed in the WHO in the last 10 years, as well as a large number of studies and articles produced by scholars, point to the shortcomings and incoherence in the current R&D model (Schumacher et al., 2016). At the end of 2015, the Secretary-General of the United Nations established a High-Level Panel on Access to Medicines; the panel was constituted by an array of personalities and international experts of demonstrated competence. The terms of reference set for the expert group called for a study on 'the incoherence between the rights of inventors, international human rights legislation, trade rules and public health' (UNHLP, 2015). As noted earlier, although an encouraging path to go to a new direction was opened in 2013 at the WHO with the recommendations of the Consultative Expert Working Group on Research and Development: Financing and Coordination (CEWG), such recommendations have not been implemented so far (World Health Assembly, 65, 2012).

4.3.4 Fragmentation and Lack of Coordination

At the time the novel coronavirus started to spread in 2020, it was clear that the stocks or production capacity of masks or alcohol-based hand rub or breathing assistance devices were unknown. Who were the producers and how could they respond to the quantities needed? Prices shot up and some countries imposed export restrictions. The European Union (EU) moved to limit exports of medical equipment outside the EU: 'We need to protect our health workers, who are in the first line of defence against the virus', said Ursula von der Leyen on 15 March 2020 (Bayer et al., 2020). What is valid for production and distribution also applies to research and development of vaccines and possible future treatments. The WHO has tried to gather information and when it has it (in case private and public actors provide it) what will it do with this information, how will the organisation be able to set the rules of the game?

The search for new treatments and health technologies – as well as the production and distribution of products necessary for the protection of life and recovery of health – should be carefully planned and subject to well defined rules. Sharing information is fundamental but it is not enough. The world is interdependent in relation to R&D for and the production of pharmaceuticals. This current crisis has dramatically shown the need for cooperation in the field of research, development and production of pharmaceuticals. Sharing of technologies, and not only information about them, is essential to maintain the supply of vital products. No country is totally self-sufficient. Closing borders and restricting exports may be a palliative, but not a solution. The only solution is a global coordination of all actors. This is a role that the WHO could play if the organisation is allowed to use the legal instruments available under its constitution.

The WHO 'R&D Blueprint is a global strategy and preparedness plan that allows the rapid activation of R&D activities during epidemics. Its aim is to fast-track the availability of effective tests, vaccines and medicines that can be used to save lives and avert large scale crisis' (WHO, 2021). This is an excellent but insufficient initiative in view of what is happening now. If the WHO has the information, it is already one step, but the information is only the basis for decision making. Who will make the decisions? And what will be the instruments for their implementation? The WHO cannot be a world health government without laws and instruments to enforce those laws. As noted by Viergever, '[o]ne of the most pressing global health problems is that there is a mismatch between the health research and development (R&D) that is needed and that which is undertaken. The dependence of health R&D on market incentives in the for-profit private sector and the lack of coordination by public and philanthropic funders on global R&D priorities have resulted in a global health R&D landscape that neglects certain products and populations and is characterized, more generally, by a distribution that is not "needs-driven"'. (Viergever, 2013).

4.3.5   *Waste and Overlap*

There is waste and overlap in vaccine and treatments research. According to information from the WHO Blueprint there is a number of research studies on the vaccine candidate (in China, Australia, the UK, Canada, France, Germany, US, etc.). As there is little or no exchange on research progress between the different countries, resources are spent looking for what others have probably already found. According to the WHO, there are currently clinical trials for 5 vaccine candidates (WHO Blueprint, 2020). Research with the same objective is done in different sites and countries. There is no information in the WHO Blueprint on whether progress is shared on different research, particularly among those working with the same platform technologies. Not sharing research results extends the time and costs of the process. In January 2020, RAND Europe wrote in its report on innovating for better healthcare: 'A variety of funding schemes support innovation in the health system, but there is a need to improve the coordination, sustainability and stability of funding flows' (Marjanovic et al., 2020).

According to the WHO Blueprint, there are several ongoing research efforts on existing drugs:

- 'In vitro studies of antiviral agents
- Cross-reactivity studies to evaluate monoclonal antibodies (mAbs) developed against SARS
- Clinical trials in China (>85)

- Remdesivir
- Lopinavir+Ritonavir
- Tenofovir, Oseltamivir, Baloxivir marboxil, Umifenovir
- Novaferon
- Interferons (IFNs)
- Chloroquine
- Traditional Chinese Medicines: Lianhua Qingwen' (WHO Blueprint, 2020)

The WHO should also ensure that all pandemic-related products (existing or to be developed) be treated as *public goods*, that is, they should be available to producers around the world to be able to respond to a massive demand, something that a single or group of producers would not be able to achieve. This should be part of an internationally agreed pandemic declaration. Some antivirals and other existing medications are being tested to see if they could be used for treatment of persons infected with the coronavirus. It is not yet clear whether there will be patents for second uses or new indications. This kind of patents is not required under the TRIPS Agreement and, if granted, they may pose important barriers to access of medicines (Ducimetière, 2019).

4.4   A Binding International Convention

As noted earlier, there is only one historical precedent for the use of Article 19 of the WHO Constitution: The Framework Convention on Tobacco Control (FCTC). It was adopted in May 2003 and has now been signed by 168 countries. For the first time, the WHO exercised the power to adopt international treaties and agreements in a substantive area and provided a global legal response to a global health threat.

The WHO Framework Convention on Tobacco Control is a framework treaty which, while alluding to many substantive issues, essentially sets out the objectives, principles, institutions, and functioning of what should be a more comprehensive system with the adoption of future additional protocols on technical issues, such as promotion and sponsorship, advertising, illicit trade, and liability (Devillier, 2005).

According to the report of the Eighth Session of the Conference of the Parties 2018 (COP8) to the WHO FCTC, Vera Luiza da Costa e Silva, Head of the WHO FCTC, said: 'We are happy to report, based on the information received from the Parties in the 2018 reporting cycle, that progress is evident in implementation of most articles to the Convention, especially the time bound measures concerning smoke-free environments, packaging and labelling and tobacco advertising, promotion and sponsorship ban' (da Costa, 2018).

The finding that the current system of incentives through the protection of patents has failed to respond to the global health problems shows the urgency of using efficient mechanisms to ensure and enable universal health coverage. The success of FCTC should serve as inspiration.

As far as sustainable long-term access to medicines for the developing countries and today even for developed countries is not available, the WHO should, rather than recommend, use its capacity to legislate: a convention or a treaty on R&D is undoubtedly one the paths to follow. As noted by the report of the WHO Commission on Intellectual Property Rights, Innovation and Public Health (CIPIH), 'there is a need for an international mechanism to increase global coordination and funding of medical R&D, the sponsors of the medical R&D treaty proposal should undertake further work to develop these ideas so that governments and policy-makers may make an informed decision' (WHO, 2006b).

4.4.1   Objective and Scope

The objectives of an international and binding treaty for R&D and innovation for health would be as follows: promote R&D for all diseases, conditions or problems (including pandemic outbreaks), promote R&D capacity in developing countries and with a sustainable and affordable model that prioritises public interest and health.

4.4.2   Possible Main Components

To achieve this goal, an international treaty must include the following:

- The establishment of priorities based on public health needs.
- Coordination of public R&D on pharmaceuticals.
- Develop sustainable financing mechanisms.

Priority setting would aim to ensure that the R&D programme in medicines and health technologies is based on the public health needs of the population and not on potential commercial benefits.

A key component of a binding global R&D treaty should be the development of R&D coordination mechanisms to achieve clearly identified objectives at the lowest possible cost. All actors (public and private) should be informed and/or guided in the allocation of resources, and R&D efforts can be monitored and evaluated. Mechanisms to be agreed upon may include the creation of networks of existing institutions, particularly in developing countries, and the creation of new pro-grammes and facilities.

A binding international R&D treaty should propose the establishment of a funding mechanism, based on the transparency of research and development costs. The source of funding for the fund would come from governments, with contributions according to their level of development (Muñoz Tellez, 2020).

4.5   The Need to Act Fast

In the face of the health crisis, in March 2020 Canada, Chile, Ecuador and Germany have taken steps to facilitate their right to issue compulsory licenses for COVID-19 present and future diagnostics, medicines, vaccines and other medical products and technologies (Muñoz Tellez, 2020). Similarly, the government of Israel issued a compulsory license for patents on a medicine they were investigating for use against COVID-19 (MSF, 2020). On 14 March, Spain issued a decree declaring the state of emergency, giving the government the power to intervene and temporarily occupy factories in the pharmaceutical sector; to enforce the orders necessary to **guarantee the supply of medicines** and products necessary for the protection of public health, and also to adopt special measures in relation to the manufacture, importation, distribution and dispensation of medicines (Lopez, 2021). Other governments have taken similar measures. These isolated and uncoordinated efforts would be more effective in the context of a global response.

A WHO declaration of pandemic should include, among other key elements, a recognition of the right of countries not to enforce exclusive rights under patents or other intellectual property rights in relation to all present and future health products (diagnostics, treatment and vaccines) related to the pandemic. In an open letter to the Director Generals of the WHO, WIPO, and WTO, the Executive Director of the South Centre stated that 'access to affordable medicines, vaccines and diagnostics and to medical equipment, and to the technologies to produce them, is indispensable to treat COVID-19' and that such technologies 'should be broadly available to manufacture and supply what is needed to address the disease. Any commercial interest supported by the possession of intellectual property rights on those technologies must not take precedence over saving lives and upholding human rights. This should always be the case, but this premise is often overlooked in times where asymmetries in development and inequality are deemed to be normal facts'. The letter also called upon the three organisations, to 'support developing and other countries, as they may need, to make use of article 73(b) of the TRIPS Agreement to suspend the enforcement of any intellectual property right (including patents, designs and trade secrets) that may pose an obstacle to the procurement or local manufacturing of the products and devices necessary to protect their populations' (South Centre, 2020).

In summary, there is a need to act fast and in a coordinated manner at the global scale. While the necessary international tools and mechanisms are not in place now, this crisis will hopefully leave a major (albeit hard learned) lesson: there is a need to rethink the R&D model as part of a new and more effective governance of global health issues.

4.6 Conclusions and Recommendations

- As a starting point, in cases such as the present COVID-19 pandemic, the WHO should include in the pandemic declaration a call for all products and technologies related to the pandemic to be treated as public goods.
- The global health crisis caused by the coronavirus COVID-19 pandemic creates an opportunity to rethink and put in place an R&D model for pharmaceutical products and health technologies that is more efficient, less costly and responsive to health needs.
- There is a need for sustainable long-term innovative mechanisms to promote pharmaceutical R&D to address public health needs, particularly in developing countries, in the context of a model that structurally links innovation with access.
- The WHO Member States should, based on Article 19 of the WHO Constitution, start negotiations for a binding global instrument for R&D and innovation for health, as recommended by the WHO-CEWG and the UN High-Level Panel on Access to Medicines.
- A successful binding global instrument for R&D must be able to prioritise R&D in accordance with health needs, to coordinate R&D to avoid unnecessary duplication of efforts and to design sustainable public mechanisms for financing for R&D. The world would be better prepared for a health crisis as the one created by the COVID-19 pandemic.
- As noted in the open letter mentioned earlier, '[we] need to have the courage to change course. The resource gap in addressing the health crisis is huge and health inequality is probably the most unbearable of injustices. It will be a matter of rebuilding a world that is viable, the one we are leaving behind, was not' (Lopez, 2021).

References

Barbancey, P. (2020). Les contraintes d'accès à un vaccin seront sévères sans une action internationale concertée. *L'Humanité*, 26 March 2020. https://www.humanite.fr/les-contraintes-dacces-un-vaccin-seront-severes-sans-une-action-internationale-concertee-686737.

Bayer, L., Deutsch, J., Hanke Vela J., & Tamma, P. (2020). EU moves to limit exports of medical equipment outside the bloc. *Político*, 15 March 2020. https://www.politico.eu/article/coronavirus-eu-limit-exports-medical-equipment/.

Carrington, D. (2020). Coronavirus: 'Nature is sending us a message', says UN environment chief. *The Guardian*, 25 March 2020. https://www.theguardian.com/world/2020/mar/25/coronavirus-nature-is-sending-us-a-message-says-un-environment-chief.

Connell, J. W., Blanckley, A., Boldon, H., & Warrington, B. (2012, March 1). Diagnosing the decline in pharmaceutical R&D efficiency. *Nature Reviews Drug Discovery, 11*(3).

da Costa, V. L. (2018). Head of the WHO/FCTC, Opening Remarks at the COP8. Geneva, 1 October 2018. https://www.who.int/fctc/secretariat/head/statements/2018/cop8-open-remarks-head-secretariat/en/.

Devillier, N. (2005). La convention-cadre pour la lutte anti-tabac. *Revue Belge du Droit International, 1–2*, 172.

DNDi. (2019). 15 years of needs driven innovation for access. Geneva.

Ducimetière, C. (2019). *Second medical use patents – Legal treatment and public health issues*, Research Paper No. 101. : South Centre, December 2019. https://www.southcentre.int/research-paper-101-december-2019/.

Jung, E. H., Engelberg, A., & Kesselheim, A. S. (2019). Do large pharma companies provide drug development innovation? Our analysis says no. December 2019, Biosimilars STAT Reports, Boston USA. https://www.statnews.com/2019/12/10/large-pharma-companies-provide-little-new-drug-development-innovation/.

Light, D. W., & Warburton, R. (2011). Demythologizing the high costs of pharmaceutical research. *BioSocieties, 6*, 34–50. https://doi.org/10.1057/biosoc.2010.40

Lopez, V. (2021). Pharmaceutical policy in times of health emergency. https://saludporderecho.org/en/pharmaceutical-policy-in-times-of-health-emergency/. Accessed 25 June 2021.

Lurie, N., Saville, M., Hatchett, R., & Jane Halton, J. (2020). Developing Covid-19 vaccines at pandemic speed. NEJM.org, 30 March 2020.

Marjanovic, S., et al. (2020). Innovating for improved healthcare: The current context and ways forward for quality and productivity in the NHS. January 2020. https://www.rand.org/randeurope/research/projects/innovation-as-a-driver-of-quality-and-productivity.html.

MSF. (2020). MSF statement, New York. 26 March 2020. https://www.msf.org/covid-19.

Muñoz Tellez, V. (2020). The COVID-19 pandemic: R&D and intellectual property management for access to diagnostics, medicines and vaccines. Policy Brief No. 73, South Centre, April 2020. https://www.southcentre.int/policy-brief-73-april-2020/.

Nicoli, V. (2020). Chomsky: "Las camas de los hospitales se han suprimido en nombre de la eficiencia". *Il Manifesto*, 22 March 2020. https://kaosenlared.net/chomsky-las-camas-de-los-hospitales-se-han-suprimido-en-nombre-de-la-eficiencia/.

Prescrire. (2017). Bilan Prescrire 2017 des médicaments: beaucoup des nouveautés sans progrès. 25 January 2017. https://www.prescrire.org/fr/3/31/53768/0/NewsDetails.aspx.

Schumacher, A., Gassmann, O., & Hinder, M. (2016). Changing R&D models in research-based pharmaceutical companies. *Journal of Translational Medicine, 14*, 105. https://doi.org/10.1186/s12967-016-0838-4

Seuba, X. (2010). *La protección de la salud ante la regulación internacional de los productos farmacéuticos*. Marcial Pons.

Smolar, P. (2020). Comment le Quai d'Orsay pense l'après-coronavirus, entre « compétition âpre » et « emprise de la Chine ». *Le Monde*, 31 March 2020. https://www.lemonde.fr/international/article/2020/03/31/coronavirus-comment-la-diplomatie-francaise-pense-le-jour-d-apres_6034979_3210.html.

South Centre. (2020). COVID-19 pandemic: Access to prevention and treatment is a matter of national and international security, Open letter from Dr. Carlos Correa, Executive Director of the South Centre, 4 April 2020. https://www.southcentre.int/wp-content/uploads/2020/04/COVID-19-Open-Letter-REV.pdf.

Tibau, A., Ocana, A., Anguera, G., et al. (2016). Oncologic Drugs Advisory Committee recommendations and approval of cancer drugs by the US Food and Drug Administration. *JAMA Oncology, 2*(6), 744–750. http://jamanetwork.com/journals/jamaoncology/article-abstract/2497879

Toebes, B. (2015). International health law: An emerging field of public international law. *Indian Journal of International Law, 55*, 299–328. https://doi.org/10.1007/s40901-016-0020-9

Tufts Center for the Study of Drug Development. (2014). Cost of developing a new drug. Boston, November 2014.

UNHLP. (2015). United Nations Secretary-General's High-Level Panel on Access to Medicines, NY. http://www.unsgaccessmeds.org.

Van Norman, G. A. (2017). Overcoming the declining trends in innovation and investment in cardiovascular therapeutics: Beyond EROOM's Law. *JACC: Basic to Translational Science, 2*(5), 613–625. https://doi.org/10.1016/j.jacbts.2017.09.002

Velásquez, G. (2019). The most expensive drug in the history of the pharmaceutical industry. SouthViews, No. 182, South Centre, 11 July 2019. https://us5.campaign-archive.com/?u=fa9cf38799136b5660f367ba6&id=5569cbd117.

Velásquez, G. (2020). *Medicines and intellectual property: 10 years of the WHO global strategy*, Research paper no. 100. : South Centre, December 2019. https://www.southcentre.int/research-paper-100-december-2019/.

Velásquez, G., & Seuba, X. (2011). *Rethinking global health: A binding convention for R&D for pharmaceutical products*, Research Paper No. 42. : South Centre, December 2011, p 8. https://www.southcentre.int/research-paper-42-december-2011/.

Viergever, R. F. (2013). The mismatch between the health research and development (R&D) that is needed and the R&D that is undertaken: An overview of the problem, the causes, and solutions. *Global Health Action, 6*, 22450. https://doi.org/10.3402/gha.v6i0.22450

WHO. (2006a). Constitution of the World Health Organization, basic documents, 45th edn, Supplement, October 2006. https://www.who.int/governance/eb/who_constitution_en.pdf.

WHO. (2006b). Public health, innovation and intellectual property rights: Report of the Commission on Intellectual Property Rights, Innovation and Public Health., p. 178. WHO. https://www.who.int/intellectualproperty/documents/thereport/ENPublicHealthReport.pdf?ua=1.

WHO. (2011). Preparación para una gripe pandémica: Marco para el intercambio de virus gripales y el acceso a las vacunas y otros beneficios. https://apps.who.int/iris/bitstream/handle/10665/44867/9789243503080_spa.pdf;jsessionid=83D46BFFE59F6AB5C5F423EF6A0D1474?sequence=1.

WHO. (2019). *Global Preparedness Monitoring Board, a world at risk: Annual report on global preparedness for health emergencies*. World Health Organization. https://apps.who.int/gpmb/assets/annual_report/GPMB_annualreport_2019.pdf

WHO. (2020). WHO Director-General's opening remarks at the media briefing on COVID-19. 18 March 2020. https://www.who.int/dg/speeches/detail/who-director-general-s-opening-remarks-at-the-media-briefing-on-covid-19%2D%2D-18-march-2020.

WHO. (2021). WHO R&D Blueprint. https://www.who.int/observatories/global-observatory-on-health-research-and-development/analyses-and-syntheses/who-r-d-blueprint/background.

WHO Blueprint. (2020). 2019 novel Coronavirus Global research and innovation forum: towards a research roadmap. https://www.who.int/blueprint/priority-diseases/key-action/Overview_of_SoA_and_outline_key_knowledge_gaps.pdf?ua=1.

World Health Assembly, 65. (2012). Consultative expert working group on research and development: Financing and coordination. WHO. https://apps.who.int/iris/handle/10665/79197.

Chapter 5
Intellectual Property and Access to Medicines and Vaccines

5.1 Introduction

Intellectual property (IP) and patents have become one of the most debated issues on access to medicines since the creation of the World Trade Organization (WTO) and the coming into force of the Agreement on Trade-Related Aspects of Intellectual Property Rights (TRIPS). Patents are by no means the only barriers to access to life-saving medicines, but they can play a significant or even determining role. During the term of patent protection, the patent holder's ability to decide on prices, in the absence of competition, can result in the medicine being unaffordable to the majority of people living in developing countries.

This chapter aims, in its first part, to introduce key aspects of access to medicines and intellectual property. The second part describes and defines some of the basic terms and concepts of the relatively new area of pharmaceutical policy, the trade-related aspects of intellectual property rights that regulate the research, development and supply of medicines and health technologies in general.

5.2 The WTO TRIPS Agreement

The World Trade Organization (WTO) is an international organisation of (currently) 164 Member States dealing with the rules of trade and providing the institutional framework for the conduct of trade relations among its members. On joining the WTO, members adhere to several agreements, and of these the Agreement on

This chapter is largely taken from: Velásquez, G. (2019 December). *Intellectual Property and Access to Medicines: An Introduction to Key Issues - Some Basic Terms and Concepts*. South Centre Training Paper 1. https://www.southcentre.int/wp-content/uploads/2019/12/TP1_Intellectual-Property-and-Access-to-Medicines_EN-1.pdf.

G. Velásquez, *Vaccines, Medicines and COVID-19*, SpringerBriefs in Public Health, https://doi.org/10.1007/978-3-030-89125-1_5

Trade-Related Aspects of Intellectual Property Rights (TRIPS) certainly has the greatest impact on the pharmaceutical sector.

The TRIPS Agreement establishes minimum standards for the protection and enforcement of a set of intellectual property rights that WTO Members are required to implement through national legislation. The TRIPS Agreement was adopted and came into force in 1995, but countries could benefit from different transition periods according to their economic development and the protection that they had granted to intellectual property until then. Before the TRIPS Agreement, patent issues were treated differently in each country and countries had different approaches to patent (and other types of intellectual property) protection to cater for their different needs.

5.3 What Is a Patent?

A patent is a title granted by the public authorities conferring temporary monopoly for the exploitation of an invention. It provides the patent holder a negative right; that is, the right to prevent others from using, making, selling, importing or marketing the patented invention during the term of the patent, without the permission or consent of the patent holder.

5.3.1 There Is No Global or International Patent

An important concept related to patent rights is **territoriality**. What this means is that the rights over a patented invention have a limited geographic coverage. In many cases, patents are granted by national patent offices, governed by the patent legislation in force in the country. The territorial reach of the patent right in such cases is national, i.e. the patent-holder of a patent granted by the patent office of Country A, will not have patent rights in Country B, unless a patent has also been similarly granted in Country B.

In some cases, there may be a regional patent office; in which case, a patent granted by the regional patent office may be recognised in the countries that are members of the regional patent agreement, subject to different conditions and procedures. For example, the European Patent Office may grant an EPO patent, which is recognised by all parties of the European Patent Convention. In this case, such a patent is regarded as a 'bundle of nationally enforceable' rights; that is to say, the rights accruing to the patent will have to be individually enforced in each member country.

The African Intellectual Property Organization, which is better known as OAPI (derived from the acronym of its name in French: *Organisation Africaine de la Propriété Intellectuelle),* is a regional patent organisation that acts as the common patent authority for the 16 OAPI Member States (i.e. Benin, Burkina Faso, Cameroon, Central African Republic, Chad, Congo, Cote d'Ivoire, Equatorial

Guinea, Gabon, Guinea, Guinea Bissau, Mali, Mauritania, Niger, Senegal and Togo). The unique feature of the OAPI patent regime is that a patent granted by OAPI will automatically apply in each of the OAPI Member States. OAPI thus functions as the national patent office for all its Member States, receiving applications and granting patents. While an application may be filed with the relevant national administration in a Member State, OAPI is the body responsible for the granting of the patent. Once granted, the rights accruing to a patent are independent of national rights, defined under the provisions of the Bangui Agreement but also subject to the national legislation, if any, of the Member States. In contrast, the African Regional Intellectual Property Organization (ARIPO) permits filing of one patent application (designating the countries in which protection is sought) at the Industrial Patent Office of any contracting state or directly with ARIPO but does not have automatic national effect in its Member States. The 16 Member States of ARIPO (Botswana, Gambia, Ghana, Kenya, Lesotho, Malawi, Mozambique, Namibia, Sierra Leone, Somalia, Sudan, Swaziland, Tanzania, Uganda, Zambia and Zimbabwe) may reject patents granted by ARIPO within 6 months of receipt of the notification, on the basis that they are contrary to national legislation or that they do not comply with the provisions of the Harare Protocol on patents, marks, models and designs (Shashikant, 2014).

5.3.2 The Patent Cooperation Treaty

The Patent Cooperation Treaty (PCT) adopted in 1970 is one of the treaties administered by the World Intellectual Property Organization (WIPO) with more than 150 contracting states. PCT makes it possible to seek patent protection for an invention simultaneously in many countries by filing a single 'international' patent application instead of filing several separate national or regional patent applications. The granting of patents remains under the control of the national or regional patent offices in what is called the 'national phase'.[1]

After 'international' patent application is filed in the patent office of a PCT member state or in the International Bureau of WIPO, a search and examination is then conducted on that application by a patent office of a PCT member state that is recognised as a PCT International Search and Examination Authority. The 'international' patent application can be filed within a period of 12 months from the first filing of the corresponding patent application in any state that is party to the Paris Convention. The International Search Authority to which the application is transmitted then conducts a prior art search based on published documents and issues a written opinion and an international search report on whether the application meets generally the criteria of patentability based on the prior art search, without any assessment of the application against national legal standards on the thresholds of

---

[1] WIPO, PCT FAQS, https://www.wipo.int/pct/en/faqs/faqs.html.

patentability criteria. The application and the written opinion and the search report are then published within a period of 18 months from the first filing of the application in any country. The applicant then has the option to request a supplementary search by another patent office recognised as an International Search Authority. The applicant also the option to request a supplementary international examination to analyse the patentability of the application, usually based on an amended version of the application. These requests can be made within a period of 22 months from the initial application. The International Preliminary Report on Patentability or the Supplementary International Search Report is issued within 22 months. Following this, the applicant can decide on whether to pursue national phase prosecution of the patent application and the request the same to the respective national offices within a period of 30 months from the initial application.

The national patent offices are not bound by the international search and examination report but may rely on it in course of their own search and examination. However, this also allows patent offices that produce the international search and examination report in their capacity as International Search Authority (ISA) to influence the national examination of that application in a developing country.[2] Indeed, as explained by the WIPO Secretariat, one advantage of the PCT system is that '… the search and examination work of patent offices can be considerably reduced or virtually eliminated ….'[3] (Syam, 2019).

The bilateral and regional free trade agreements promoted by the United States and the European Union (EU), typically introduce an obligation for developing countries to join PCT. According to Syam (2019), 'while a large number of developing countries have acceded to the PCT, the system is predominantly used by applicants from a few countries. Many developing countries that have joined the PCT system lack capacity in conducting substantive examination, though they have witnessed significant increase in the number of patent applications filed in their countries through the PCT route'.

5.3.3 Validity of Patents

The fact that a patent has been granted by a patent office does not mean that this is the final say on the matter. A granted patent can sometimes be partly or completely invalidated, for a number of reasons. For example, if on closer scrutiny, it is found that the patent does not meet one or more of the patentability criteria (as set out in the national patent law); it may be possible to challenge its validity.

Patent laws may also have provisions that exclude certain kinds of inventions: common examples are therapeutic or surgical methods. Patent laws may also

---

[2] See Peter Drahos, 'Trust Me: Patent Offices in Developing Countries', *American Journal of Law & Medicine*, vol. 34 (2008), pp. 151–174.

[3] Available from http://www.wipo.int/pct/en/treaty/about.html.

exclude the patenting of inventions when their commercialisation is prohibited because the invention would be contrary to *ordre public* or morality. Patents granted in the excluded fields would also be invalid.

Even where a patent has been properly granted, the patent holder must maintain the patent by paying the required maintenance fees to the patent office. When the fees are not paid, the patent will lapse and therefore will no longer be valid.

5.3.4   Minimum Standards of Patent Protection

The minimum standards that the TRIPS Agreement requires for the protection of patent rights include the following:

- All WTO members have to provide patent protection for 'inventions', in all fields of technology. In the case of pharmaceuticals, WTO members have to grant patents to any invention of pharmaceutical product or process.
- WTO members shall apply the patentability criteria of novelty, inventive step (non-obviousness), and industrial application (utility). However, there is room for individual countries to determine the actual definition and application of these criteria.
- The fact that the TRIPS Agreement does not define novelty, inventive step and industrial applicability leaves countries significant room for manoeuvre; therefore, patentability requirements represent the principal and most important flexibility allowed by the Agreement to protect public health and access to medicines (Velásquez, 2015). 'Politicians and legislators have broad room for manoeuvre to give legal effect to those flexibilities' (Arias, 2014).
- The TRIPS Agreement also requires a minimum term of protection for patent rights of 20 years from the date of filing the application. Thus, WTO members cannot now have a shorter duration of patent protection than the minimum required 20 years.

Although the minimum duration required by the TRIPS Agreement is 20 years, a report from I-MAK, analyses the 12 best-selling drugs in the United States and reveals that drug makers file a large number of patent applications to extend their

Table 5.1 Examples of drugs with multiple patents granted

Product	Company	Conditions treated	No. patents granted	Years of protection
Humira	ABBVIE	Arthritis	132	39
Rituxan	BIOGEN	Cancer	94	47
Revlimid	CELGENE	M. Myeloma	96	40
Enbrel	AMGEN	Arthritis	41	39
Herceptin	ROCHE	Cancer	108	48

Source: I-MAK 'Overpatented, Overpriced', Nov. 2018

monopolies far beyond the 20 years of protection intended under patent law. Some examples are shown in Table 5.1 (I-MAK, 2018).

However, the TRIPS Agreement did not impose a uniform international law or uniform legal requirements. It contains provisions that allow for a degree of flexibility and some room for countries to accommodate their own patent and intellectual property systems according to their developmental needs. Thus, WTO Members are still able to determine how certain aspects of patent protection may be applied or implemented at the national level, in accordance with the social and economic welfare of the country.

Article 7 of the TRIPS Agreement, which spells out the 'Objectives' of the Agreement, provides that protection of intellectual property rights: 'should contribute to the promotion of technological innovation and to the transfer and dissemination of technology, to the mutual advantage of producers and users of technological knowledge and in a manner conducive to social and economic welfare'. In addition, WTO Members are allowed to 'adopt measures necessary to protect public health and nutrition, and to promote the public interest in sectors of vital importance to their socio-economic and technological development ...', as stated in Article 8, which lays down the Principles of the TRIPS Agreement (WTO, 1994).

These two provisions, together with the Preamble of the TRIPS Agreement, reflect the fundamental tenet that intellectual property rights protection should be regarded as a public policy tool; that is to say, the protection of such rights should be balanced against other public interests to achieve public policy goals.

5.3.5   Patents on Pharmaceutical Products

The conventional rationale for patent protection can be explained as follows: by conferring a temporary or time-limited monopoly, patents allow the inventor/producer to recover the costs of investment in research and development, and also to earn a profit in the production and sale of the invention. This is in return for making publicly available the knowledge about the invention, so that further research and development, and subsequent innovations, can be stimulated. Therefore, patent protection can be seen as a bargain struck by society with the patent holder, based on the premise that without patent protection there would be insufficient incentive for innovation. It is also based on the assumption that consumers would be better off in the long term because the short-term cost of having to pay higher prices will be offset by the creation of new inventions thanks to additional research and development.

However, questions arise as to whether these assumptions are always borne out in practice. In the area of public health and patents on pharmaceuticals, these questions have been particularly persistent.

In the case of pharmaceuticals, it is argued that patents are crucial for pharmaceutical innovation, and that without patent protection, there would be no financial incentive to fund the costs of discovery and development of new medicines. It is true

that patent protection has provided an important incentive mechanism to drive research and development in the pharmaceutical industry. Yet it is also true that patented medicines are normally priced well above production costs to obtain significant profits after paying marketing costs that frequently surpass those of research and development (Washington Post, 2015). In some developing countries, the high price of certain medicines means that patients in these countries will not have access to treatment.

Developing countries account for a very small fraction of the global pharmaceutical market (USA, EU and Japan accounted in 2018 for 89.3% of world pharmaceutical sales) and the generation of income to fund more research and development is not dependent on the profits derived from their markets (EFPIA, 2018). Indeed, the patent protection system has provided little incentive for research and development of new medicines needed for diseases afflicting developing countries (Pedrique et al., 2013). This highlights some of the difficulties in relying solely on patent protection as the incentive system and on the private sector to develop essential medicines. The WHO Commission on Intellectual Property Rights, Innovation and Public Health (CIPIH) was tasked with analysing these issues, amongst others. In its Report, CIPIH stated that 'because market demand for diagnostics, vaccines and medicines needed to address health problems mainly affecting developing countries is small and uncertain, the incentive effect of IPRs may be limited or non-existent' (WHO, 2006). Thus, there is a need for other incentives and financial mechanisms to be put in place, which is what the WHO Global Strategy and plan of action (GSPOA) on Public Health, Innovation and Intellectual Property refers to in the WHA Resolution 61.21.

Another concern relates to the subject matter and number of patents that are granted to provide protection for pharmaceutical products. While only a small number of new chemical entities are approved annually, many patent applications for the protection of pharmaceutical products are submitted. For example, the number of new molecular entities (NMEs) approved by the US Food and Drug Administration has drastically declined since the mid-1990s (from 53 in 1996 to 22 in 2016) (Nature Reviews Drug Discovery, 2019). Patent applications for pharmaceuticals are not filed merely on the newly discovered chemical molecule or compound. Patents have increasingly been filed and often granted on variants of a pharmaceutical product, such as salts and other derivatives of the molecule and the specific formulation or dosage form of the medicine. Even so-called 'incremental' modifications of existing products, including slight modifications or trivial features such as the form, colour and inert ingredients, have been claimed and obtained patent protection in some countries. Patents have also been granted on the combinations of a known medicine with other known medicines. The granting of these various patents means that a particular pharmaceutical product may be protected during many years although the patent on the chemical molecule on which it is based has expired.

In these circumstances, the criteria applied to examine and grant pharmaceutical patents are extremely relevant for public health policies and not only a matter of concern for patent and industrial policy. Policy makers in the public health area, as well as patent examiners, should be aware that decisions relating to the granting of a patent can directly and unduly affect the health and lives of people.

5.3.6   Patents and Access to Essential Medicines

The HIV/AIDS pandemic and the urgent need to make treatment available for the 14.6 million people in need of treatment (at the end of 2018) continue to bring the question of the affordability of antiretroviral (ARV) medicines to the forefront of international attention (UNAIDS, 2019).

When ARVs were first introduced, the cost of treatment per person was over US$ 10,000 a year (about US$ 30 a day). This cost put ARVs out of reach for the vast majority of HIV patients in developing countries, where more than 3 billion people live on less than US$ 2 a day.[4] Introduction of competition has resulted in significant reductions in the prices of ARVs. Since then, there has been an increasing reliance on low-cost generic ARV therapy as a strategy for treating more patients; today the annual first line treatment per person is available at less than US$ 100.

HIV/AIDS was one of the detonating factors of the controversy on patents and access to medicines. Affordability of treatment for other diseases affecting millions of people, such as hepatitis C, malaria, diabetes, cancer, tuberculosis or cardiovascular diseases is also now part of the debate.

5.4   The Doha Declaration on the TRIPS Agreement and Public Health

Although the TRIPS Agreement has introduced a multilateral framework with minimum binding standards for the protection of intellectual property rights, there still exists flexibility within the provisions of the Agreement that permits countries to determine how intellectual property rules should be interpreted and applied to make them more consistent with their national public interest and priorities. However, some governments have been unsure of how that flexibility would be interpreted and how far their rights to use it would be respected.

Although TRIPS affords some discretion about how its obligations are interpreted and implemented by national governments, developing countries have faced obstacles when they sought to use measures to promote access to affordable medicines. For example, when the South African Medicines and Related Substances Act was amended in 1997 to enable parallel importation, the provision was challenged by 39 pharmaceutical companies and the South African Pharmaceutical Manufacturers' Association (PMA) before the Supreme Court of South Africa. The pharmaceutical companies eventually withdrew their legal suit because of a strong reaction from international organisations (notably WHO) and civil society. In another case, the United States challenged the legality of the Brazilian legislation that authorises the grant of compulsory licences in cases where the patent holder has

---

[4]DO Something.org. '11 Facts About Global Poverty'. https://www.dosomething.org/us/facts/11-facts-about-global-poverty.

not 'worked' their invention locally (i.e. to manufacture the patented product in the country). The US Government initiated a complaint under the WTO dispute settlement system against Brazil but later withdrew its complaint in 2001.

Other examples are referred to in the report of the United Nations Secretary-General's (UN SG) High-Level Panel on Access to Medicines such as Thailand's 2006 decision to import generic versions of the antiretroviral medicine efavirenz from India under compulsory licence (UNHLP, 2016). This decision was met with hostility from the manufacturer, Merck, and the United States Government, which questioned the legality of the compulsory licence and pressed Thailand to rescind its decision. Thailand's subsequent decision to issue two further compulsory licences in 2007 for lopinavir/ritonavir and clopidogrel also resulted in retaliatory measures. Abbott withdrew from the Thai market all medicines awaiting registration in the country. The European Trade Commissioner wrote to the Thailand Government criticising its use of compulsory licences as 'detrimental' to medical innovation, noting that such approaches could lead to Thailand's isolation from the global biotechnology investment community and urging negotiations (UNHLP, 2016).

In early 2016, the Ministry of Health of Colombia adopted resolution 2475, declaring that access to imatinib, a medicine that appears on the WHO Essential Medicines List, was of 'public interest' for the treatment of leukaemia. The resolution was a legal step necessary for the subsequent issuance of a compulsory licence. Letters sent to the co-chair of the UN SG report on Access to Medicines chronicle attempts by various domestic and foreign parties to dissuade the Colombian Government from issuing a compulsory licence as allowed by the TRIPS Agreement and the Doha Declaration.

Paragraph 4 of the Doha Declaration provides important guidance on the interpretation and implementation of the TRIPS Agreement, setting out the basic principle as follows:

> We agree that the TRIPS Agreement **does not and should not** prevent Members from taking measures to protect public health. Accordingly, while reiterating our commitment to the TRIPS Agreement, we affirm that the Agreement **can and should be** interpreted and implemented in a manner supportive of WTO Members' right to protect public health and, in particular, to promote access to medicines for all. In this connection, we reaffirm the right of WTO Members to use, to the full, the provisions of the TRIPS Agreement, which provide flexibility for this purpose. (Emphasis added.) (WTO, 2001)

5.5 What Are the TRIPS Flexibilities?

The resolution (WHA49.14) on 'Revised Drug Strategy' requested the WHO Director-General to undertake a study on the impact of the WTO, and particularly the TRIPS agreement, on access to health. This study was entrusted to the WHO Drugs Action Programme (DAP). In November 1997, the DAP published the study 'Globalization and Access to Drugs: Perspectives on the WTO TRIPS Agreement', commonly known in the WHO as the 'red book' on the TRIPS Agreement (Velásquez & Boulet, 1997, p. 58).

The WHO 'red book' speaks about 'margins of freedom'[5] (Velásquez & Boulet, 1997, p. 34). Subsequently, in March 2001, the WHO adopted the term 'safeguards' in a widely distributed document available in the six WHO official languages (WHO, 2001). In June 2001, the European Commission talks about 'a sufficiently wide margin of discretion' regarding the implementation of the TRIPS Agreement (European Commission, 2001, p. 1). A few months later, in November 2001, the Doha Declaration on the TRIPS Agreement and Public Health referred to 'the provisions of the TRIPS Agreement that provide flexibility'.[6] It is only in June 2002 that WHO referred to TRIPS 'flexibilities', in a paper analysing the implications of the Doha declaration, authored by Carlos Correa (WHO, 2002, p. 13) (emphasis added).

The Doha Declaration confirmed that the TRIPS Agreement permits governments to consider and implement a range of options that take public health into account, when formulating intellectual property laws and policy, at national and regional levels. It specifically referred to several aspects of flexibility within the TRIPS Agreement, including the right to grant compulsory licences and to permit parallel importation. This means that countries cannot be prevented from taking certain measures that limit exclusive patent rights, where the interests of public health and the need to ensure access to affordable medicines so require.

The main public health-related flexibilities available under the TRIPS Agreement are briefly described below.

5.5.1   Criteria for Patentability

A patent is granted when the application satisfies the criteria for patentability, as laid down in the national (or regional) patent legislation. According to Article 27 of the TRIPS Agreement (WTO, 1994) all national legislations must require a patent application to satisfy the three-fold criteria of:

- **novelty** – the invention must be new, in that it does not form part of the current state of the art in the particular technical field or technology; the state of the art comprises everything that before the application date has been available to the public, nationally or internationally, through its description, utilisation or any other way.
- **inventive step (non-obviousness)** – the invention must not be evident for a 'person skilled in the art' (a person trained and experienced in the particular field or technology) in the light of the current state of art; and
- **industrial applicability** (utility) – the invention must be capable of being manufactured or otherwise industrially used, since the aim of the patent law is to protect technical solutions to a given problem, not abstract knowledge.

---

[5] Emphasis added.
[6] WTO. 'Doha declaration on the TRIPS Agreement and Public health', WT/MIN(01)/DEC/W/2, p. 1. Emphasis added.

The way in which the patentability criteria are applied has changed over time and across countries, depending on how governments have determined the appropriate balance of public and private interests. Although the WTO TRIPS Agreement sets out the patentability criteria, it does not provide specific directions or definitions for how these criteria should be interpreted or applied at national level. Hence, WTO members retain the ability to define and apply the criteria, as it best suits the public interest. In this context the definition and interpretation of the three criteria for patentability are probably the most important flexibility contained in the TRIPS Agreement (Correa, 2007; Velásquez, 2015).

5.5.2   Compulsory Licences

The patent holder is free to exploit the patent-protected invention or to authorise another person to exploit it. However, when reasons of public interest or the need to correct anticompetitive practices justify it, the government may allow a third party to use the invention, without the patent holder's consent, under a compulsory licence. The patent holder is therefore forced to tolerate the exploitation of his invention by a third person or by the government itself. In these cases, the public interest in ensuring broader access to the patented invention is deemed more important than the private interest of the patent holder in fully exploiting his exclusive rights. Compulsory licences thus permit third parties to use an invention, without the patent holder's consent. For example, where particular medicines are patent protected and priced out of reach of the local population, local pharmaceutical companies may obtain compulsory licences to produce generic versions of patented medicines, or to import generic versions of medicines from foreign manufacturers. There have been 108 attempts to issue compulsory licensing for 40 pharmaceuticals in 27 countries since 1995 (Son & Lee, 2018).

Compulsory licenses have been issued in developing as well as developed countries. For instance, in July 2017, the German Federal Court announced that it had affirmed the decision of the Federal Patent Court the previous year to issue a compulsory license for the HIV drug raltegravir (marketed as Isentress) (Teschemacher, 2018). Thailand issued a compulsory licence for efavirenz, an HIV/AIDS drug, and in January 2007 issued another two compulsory licences for a heart-disease medicine and for another HIV/AIDS medicine. In May 2007, Brazil also issued a compulsory licence for efavirenz.

5.5.3   Government Use

Most patent laws allow the government (or authorised agents of the government) to use privately owned patents for public, non-commercial purposes, without the consent of the patent holder. The right of the government to use a patent for public and non-commercial use is often framed in broad terms in national laws and very often the process is procedurally much simpler. In other words, it allows for the government use of patents to be 'fast-tracked', which is of importance when life-saving medicines are required urgently. There is only an obligation to inform the patent holder of the proposed use of the patent, or promptly after such use. Government use permits the public sector's production or the importation of generics, for instance, for use in public hospitals (see Box 5.1).

> **Box 5.1: Examples of Government Use**
> In October 2003, Malaysia allowed the import of generic didanosine, zidovudine and the lamivudine+zidovudine combination from India, to supply its public hospitals, under the government use provision in its Patent Law. In 2004, Indonesia authorised government use of patents to enable local production of nevirapine and lamivudine. In September 2017 Malaysia issued a 'government use' licence for sofosbuvir to treat hepatitis C.

5.5.4 Parallel Imports

Patented products that have been legitimately put on the market of the exporting country may be imported into a country without the consent of the patent holder under the principle of exhaustion of rights. This principle means that the rights holder's control over the pharmaceutical product ceases when the said product is placed in the market for the first time. Since some patented products are sold at different prices in different markets, the rationale for parallel importation is to enable the import of patented products from countries where they are sold at lower prices. For example, where the national law provides for it, there can be export of a patented medicine from Country A (where it is sold at a lower price) for sale in Country B, subject to the drug regulatory requirements of Country B. 'Developing countries were keen to clarify in the Doha Declaration, the Members' right to adopt an international principle of exhaustion of rights' (Correa, 2016).

5.5.5 Exceptions to Patent Rights

All national patent laws have provisions relating to exceptions to the exclusive rights granted by a patent (not to be confused with the exceptions to patentability), although the scope and content of these provisions vary from country to country. Exceptions to the exclusive rights granted by patents are justified on the grounds that in certain circumstances limited exercise of the patent rights is required to achieve public policy purposes of encouraging innovation, promoting education and protecting other public interests. In the context of public health, exceptions to patent rights may be extremely important in facilitating the transfer and diffusion of technologies and in facilitating the production of generic medicines. National legislation may include different types of exceptions to patent rights; the most important among them being exceptions granted for research and the so called 'early working' exception. The 'early working' exception (also known as the 'Bolar' exception) permits the production of samples of a patented medicine for the purposes of testing and approval before the end of the patent term, to enable speedy introduction of a generic product once a patent expires.

5.5.6 Flexibility in Test Data Protection

The TRIPS Agreement (WTO, 1994, Article 39.3) requires WTO Members to protect test data against unfair competition, which does not create exclusive rights. A correct interpretation and implementation of that provision avoids the burden of creating a 'data exclusivity' problematic layer of protection in addition to patent rights on pharmaceuticals. In effect, WTO Members are not obligated under Article 39.3 to confer exclusive rights on the originator marketing approval data (Correa, 2016, p. 62).

5.5.7 Avoidance of TRIPS-Plus Provisions and Policies, Including Extension of Patent Term, Data Exclusivity, Second-Use Patents, Border Measures

TRIPS-plus provisions in free trade agreements (FTAs) (or resulting from accession to WTO) may negatively affect access to medicines. Negotiators of these agreements need timely and evidence-based information to avoid, as far as possible, provisions of this kind that may reduce the accessibility and affordability of medicines through the extension (beyond 20 years) of the term of a patent, exclusive rights in respect of the results of clinical trials (data exclusivity), overbroad border measures (e.g. covering medicines in transit) and other measures affecting market dynamics (see Sect. 5.7 of this book).

5.5.8 Mitigating Implementation or Effects of TRIPS-Plus Provisions

If TRIPS-plus provisions have been accepted, however, there is a range of conditions and safeguards that may be introduced to limit the possible negative impact of such provisions, such as exceptions to data exclusivity (for instance, when a compulsory license has been granted) and limitations to the scope and length of patent term extensions.

5.5.9 Exemption for LDCs

Least developed countries (LDCs) need not grant patents for pharmaceuticals at least until 2033 (WTO, 2001). To use this policy space, some LDCs that provide for the grant of such patents would need to review their legislation or to adopt other measures to protect the government and private parties from infringement claims. They should also preserve that policy space in negotiations of free trade and other international agreements.

5.5.10 Pre- and Post-patent Grant Opposition

Procedures before many patent offices, including the United States Patent and Trademark Office (USPTO) and the European Patent Office (EPO), provide for the possibility for third parties to contribute to the examination process through 'observations' or 'oppositions' whether before or after the grant of a patent, or both. The correct implementation of these procedures helps to improve the quality of patents granted and to avoid the creation of unjustified market barriers.

5.5.11 Use of Competition Law to Address the Misuse of Patents

Competition law may be applied to correct market distortions created through the abuse of intellectual property rights. There are national precedents that may provide useful examples of best practices (UNDP, 2014). Guidelines for the competent authorities on intellectual property and competition law may be developed to facilitate the intervention of such authorities when needed to address anti-competitive practices.

5.5.12 Disclosure Requirement, Particularly for Biologics

The full and precise disclosure of an invention is crucial for the patent system to perform its informational function. Deficient disclosure may unjustifiably extend the coverage of a patent and prevent legitimate acts by third parties. This is particularly relevant for biologicals, which cannot be described in the same way as medicines produced by chemical synthesis.

5.5.13   Flexibilities in Enforcement of IP

Measures to enforce IP – such as reversal of the burden of proof, determination of damages, border measures – if overly broad, may distort competition by discouraging or preventing market entry and the availability of generic medicines. However, there is room to design such measures in a manner that is fair and equitable to all parties engaged in administrative or judicial procedures regarding IP.

5.6   The Paragraph 6 Problem and Its Solution

The so-called 'Paragraph 6' mechanism of the Doha Declaration, as implemented by the WTO Decision of 30 August 2003, was a mandate of the WTO Ministerial Conference in Doha (2001) to solve, in an 'ad hoc' manner, a problem that affected the poorest countries.

What was (is) the problem? In paragraph (f) of Article 31 of the TRIPS Agreement, it is stated that a compulsory license 'shall be authorized predominantly for the supply of the domestic market' (WTO, 1994). This limits the volume of medicines that can be exported when their production has been enabled by a compulsory license. Such provisions affect mainly those countries that lack the manufacturing capacity to produce medicines, such as the least developed countries. This is the reason why Paragraph 6 of the Doha Declaration gives a mandate to find an 'expeditious solution' to this problem (Velásquez, 2017, p. 7; WTO, 2001).

The WTO Members first agreed on a temporary solution with the General Council Decision on the Implementation of Paragraph 6 of the Doha Declaration on the TRIPS Agreement and Public Health of 30 August 2003. On 6 December 2005, WTO Members agreed to convert the waiver into a permanent solution, which would take the form of an amendment to the TRIPS Agreement. The amendment only came into force on 23 January 2017, when two-thirds of the WTO Members ratified it, although the scheduled deadline to formally accept the amendment was originally fixed for 1 December 2007. The 'solution' requested by the Doha Declaration took more than 10 years to be incorporated into the WTO rules.

The decision on Paragraph 6 contains several cumbersome conditions to ensure that beneficiary countries can import generic medicines. In 15 years only one country, Rwanda has used it once, with an importation of antiretroviral medicines from Canada. The manager of the Canadian generic firm stated after the exportation that the system was so complicated that his firm had no intention of using it again (South Centre, 2011).

One of the recommendations of the UN Secretary-General's High-Level Panel on Access to Medicines state that 'WTO Member States should review the decision in Paragraph 6 to find a solution that would allow for a quick and convenient export of pharmaceutical products produced under a compulsory license. WTO Member States should, as appropriate, adopt an exception and a permanent reform of the TRIPS Agreement' (UNHLP, 2016, p. 27).

5.7   Impact of 'TRIPS-Plus' and 'TRIPS Extra' Provisions

Several bilateral and multilateral international trade and investment agreements require countries to adopt TRIPS-plus or TRIPS extra measures. Such provisions are known as 'TRIPS-plus'.

While TRIPS-plus and TRIPS extra provisions that have been enacted unilaterally (i.e. where a country has adopted TRIPS-plus or TRIPS extra provisions on its own) may be changed where they are deemed to be inconsistent with the national public health interest, TRIPS-plus obligations entered into under bilateral and other agreements are not as easily reversed without costs. In exchange for the promise of greater access to developed country markets, a number of developing countries have accepted such TRIPS-plus or TRIPS extra obligations. These provisions have raised questions regarding their potential to compromise the use of the TRIPS flexibilities for public health purposes and for promoting innovation with respect to diseases that disproportionately affect developing country populations. The proliferation of bilateral and regional free trade agreements has increased concerns about the impact of trade agreements on access to medicines.

The World Health Assembly, in 2004, passed a resolution urging Member States to 'take into account in bilateral trade agreements the flexibilities contained in the Agreement on Trade-related Aspects of Intellectual Property Rights and recognized by the Declaration on the TRIPS Agreement and Public Health adopted by the WTO Ministerial Conference' (WHO, 2004). The need to consider the Doha Declaration and the public health-oriented flexibilities while subscribing trade agreements has been further reiterated by World Health Assembly resolutions. Similarly, the United Nations Secretary-General's High-Level Panel on Access to Medicines (2016) recommended that: 'Governments engaged in bilateral and regional trade and investment treaties should ensure that these agreements do not include provisions that interfere with their obligations to fulfil the right to health. As a first step, they must undertake public health impact assessments. These impact assessments should verify that the increased trade and economic benefits are not endangering or impeding the human rights and public health obligations of the nation and its people before entering into commitments. Such assessments should inform negotiations, be conducted transparently and made publicly available' (UNHLP, 2016, p. 28).

Some key examples of TRIPS-plus and TRIPS extra provisions are described next.

5.7.1   Extension of Patent Protection Beyond the TRIPS Minimum

The TRIPS Agreement requires a minimum patent term of 20 years from the date of filing. This patent term has been extended by provisions in certain bilateral trade agreements to compensate patent holders for any 'unreasonable delays' in the granting of the patent or unreasonable curtailment of the patent term because of the marketing approval process. No such requirement exists under the TRIPS Agreement.

5.7.2 Restrictions on the Use of Compulsory Licences

A few free trade agreements include provisions that restrict use of compulsory licences to cases of emergencies, public non-commercial use or to remedy anti-competitive practices. Such limitations are contrary to the broad discretion governments have in the granting of compulsory licences, as affirmed by the Doha Declaration.

5.7.3 Data Exclusivity

Provisions in several bilateral agreements prohibit the use of test data submitted by originator companies for obtaining marketing approval of a product to facilitate the marketing approval of the generic versions of the originator product for a certain period. Several bilateral trade agreements require a 5-year period during which such data exclusivity will prevent drug regulatory authorities from relying on submitted test data to approve generic entrants. Data exclusivity is not a requirement of the TRIPS Agreement and creates a potential barrier for generic entrants, even when there is no patent on the product. Data exclusivity may also prevent effective use of a compulsory license, in that it may not be possible to obtain marketing approval for a medicine produced or imported under compulsory licence. Furthermore, should generic manufacturers decide to produce such data, it would result in economic waste and in unethical repetition of tests for which the outcomes are already known.

5.7.4 Marketing Approval and Patent Term Linkage

Several bilateral trade agreements have included provisions that prevent national drug regulatory authorities from granting marketing approval for generic pharmaceutical products without 'consent or acquiescence' of the patent holder, when there is a relevant patent in force. This 'linkage' between the patent protection and marketing approval may prevent approvals for generic products during the lifetime of a patent, whereas the TRIPS Agreement permits generic producers to seek regulatory approval during the life of a patent without conditions. Additionally, it obliges an already overloaded national drug authority to undertake a job beyond its field of expertise and competence. In addition, commonly there are many 'secondary' patents in relation to a single drug, which may be unduly used to prevent generic competition, even when the patent on the active ingredient has expired.

5.8 Conclusions

Notwithstanding the Doha Declaration and Article 31bis of the TRIPS Agreement, there remain major challenges in the future scenario for access to medicines. Their success in securing effective access to medicines in developing countries – depends on how countries will implement intellectual property rules to optimise the TRIPS flexibilities in their national laws and whether the necessary policy decisions and measures will be taken. Major challenges for access to medicines in the context of intellectual property rights and trade agreements still exist.

Many developing countries have yet to incorporate the full range of the TRIPS flexibilities within their national laws. There may be several reasons for this delay. First, there may be a need for specific legal expertise to craft and formulate patent laws and regulations that can consider the needs and concerns of developing countries. Second, governments may be subject to pressure from the industry or other governments not to incorporate such flexibilities.

References

Arias, E. (2014). Presentation on guidelines for the examination of patentability of chemical-pharmaceutical inventions, INPI, Argentina.

Correa, C. M. (2007). *Guidelines for examination of pharmaceutical patents: Developing a public health perspective* – A Working Paper. : WHO, ICTSD, UNCTAD.

Correa, C. M. (2016). *Public health perspective on intellectual property and access to medicines – A compilation of studies prepared for WHO* (p. 110). South Centre.

EFPIA. (2018). The pharmaceutical industry figures, Key Data. https://www.efpia.eu/media/361960/efpia-pharmafigures2018_v07-hq.pdf.

European Commission. (2001). Submission to TRIPS Council (IP/C/W/280), 12 June 2001.

I-MAK. (2018). Overpatented, overpriced: How excessive pharmaceutical patenting is extending monopolies and driving up drug prices, November 2018. https://www.i-mak.org/wp-content/uploads/2018/08/I-MAK-Overpatented-Overpriced-Report.pdf.

Nature Reviews Drug Discovery. (2019). 2018 FDA drug approvals. *Nature Reviews Drug Discovery, 18*, 85–89. https://www.nature.com/articles/d41573-019-00014-x.

Pedrique, B., Strub-Wourgaft, N., Some, C., Olliaro, P., Trouiller, P., et al. (2013). The drug and vaccine landscape for neglected diseases (2000-11): A systematic assessment. *The Lancet Global Health, 1*(6), E3371–EE379. https://doi.org/10.1016/S2214-109X(13)70078-0

Shashikant, S. (2014). *The African Regional Intellectual Property Organization (ARIPO) protocol on patents: Implications for access to medicines.* Research Paper No. 56. : South Centre, November 2014. https://www.southcentre.int/wp-content/uploads/2014/11/RP56_The-ARIPO-Protocol-on-Patents_ENRev.pdf.

Son, K.-B., & Lee, T.-J. (2018). Compulsory licensing of pharmaceuticals reconsidered: Current situation and implications for access to medicines. *Global Public Health, 13*(10), 1430.

South Centre. (2011). The Doha declaration on TRIPS and public health: Ten years later – The state of implementation. Policy Brief No. 7, November 2011. https://www.southcentre.int/wp-content/uploads/2013/06/PB7_-Doha-Declaration-on-TRIPS-and-Health_-EN.pdf.

Syam, N. (2019). *Mainstreaming or dilution? Intellectual property and development in WIPO.* Research Paper No. 95. : South Centre, July 2019. https://www.southcentre.int/wp-content/uploads/2019/07/RP95_Mainstreaming-or-Dilution-Intellectual-Property-and-Development-in-WIPO_EN.pdf.

Teschemacher, R. (2018). German Federal Court of Justice confirms the compulsory license granted by way of a preliminary injunction for the AIDS drug Isentress. Bardehle Pagenberg, January 2018. http://www.mondaq.com/germany/x/667848/Patent/German+Federal+Court+o f+Justice+confirms+the+compulsory+license+granted+by+way+of+a+preliminary+injuncti on+for+the+AIDS+drug+Isentress+the+EPO+Board+of+Appeal+then+revokes+the+Europe an+patent.

UNAIDS. (2019). UNAIDS Data. https://www.unaids.org/en/resources/documents/2019/2019-UNAIDS-data.

UNDP. (2014). Using competition law to promote access to health technologies N.Y. https:// hivlawcommission.org/wp-content/uploads/2017/06/UNDP-Using-Competition-Law-to-Promote-Access-to-Medicine-05-14-2014-1.pdf.

UNHLP. (2016). Report of the United Nations Secretary-General's High-Level Panel on Access to Medicines: Promoting innovation and access to health technologies, September 2016. https://static1.squarespace.com/static/562094dee4b0d00c1a3ef761/t/57d9c6ebf5e231b2f02 cd3d4/1473890031320/UNSG+HLP+Report+FINAL+12+Sept+2016.pdf.

Velásquez, G. (2015). Guidelines on patentability and access to medicines. Research Paper No. 61, p. 22. : South Centre, March 2015. https://www.southcentre.int/wp-content/uploads/2015/03/ RP61_Guidelines-on-Patentability-and-A2M_rev2_EN.pdf.

Velásquez, G. (2017). *Intellectual property, public health and access to medicines in international organizations*. Research Paper No. 78, p. 7. : South Centre, July 2017.

Velásquez, G., & Boulet, P. (1997). Globalization and access to drugs: Perspectives on the WTO TRIPS Agreement, p. 58. WHO/DAP/98.9, Geneva, November 1997.

Washington Post. (2015). Big pharmaceutical companies are spending far more on marketing than research. 11 February 2015. https://www.washingtonpost.com/news/wonk/ wp/2015/02/11/big-pharmaceutical-companies-are-spending-far-more-on-marketing-than-research/?noredirect=on.

WHO. (2001). Globalization, TRIPS and access to pharmaceuticals. World Health Organization. WHO policy perspectives on medicines, no. 3, p. 5. https://apps.who.int/iris/ handle/10665/66723.

WHO. (2002). Implications of the Doha Declaration on the TRIPS agreement and public health, Carlos M. Correa. World Health Organization.

WHO. (2004). WHA Resolution 57.14, Scaling up treatment and care within a coordinated and comprehensive response to HIV/AIDS. WHO. https://apps.who.int/gb/ebwha/pdf_files/ WHA57/A57_R14-en.pdf.

WHO. (2006). Public health, innovation and intellectual property rights: Report of the Commission on Intellectual Property Rights, Innovation and Public Health, p. 178. WHO.

WTO. (1994). TRIPS: Agreement on Trade-Related Aspects of Intellectual Property Rights, Apr. 15, 1994, Marrakesh Agreement Establishing the World Trade Organization, Annex 1C, 1869 U.N.T.S. 299, 33 I.L.M. 1197 [hereinafter TRIPS Agreement]. https://www.wto.org/english/ docs_e/legal_e/27-trips_01_e.htm.

WTO. (2001). Doha declaration on the TRIPS Agreement and Public health. WT/MIN(01)/ DEC/W/2. https://docs.wto.org/dol2fe/Pages/FE_Search/FE_S_S009-DP.aspx?language=E& CatalogueIdList=35772,37509,46740&CurrentCatalogueIdIndex=2&FullTextHash=&HasEn glishRecord=True&HasFrenchRecord=True&HasSpanishRecord=True.

The opinions expressed in this chapter are those of the author(s) and do not necessarily reflect the views of the SC: South Centre, its Board of Directors, or the countries they represent.

Chapter 6
The World Health Organization Reforms in the Time of COVID-19

6.1 Introduction

The World Health Organization (WHO) has undergone many reforms and attempts at reform since its creation in 1948. These reforms have been largely driven by various Directors-Generals who, throughout the existence of the WHO, have sought to leave a mark on the achievements of its administration.

The reform under discussion in 2020 has been prompted by the unprecedented health crisis caused by the COVID-19 pandemic. The international community has acknowledged the legal and financial structural inadequacies of the WHO to meet its expectations.

Since the creation of the WHO, its Member States have not always been supportive of the Organization. At different times in its history, some countries have weakened it, rather than strengthened it.

In 1986, Jonathan Mann, Director of the WHO Global Programme on AIDS (GPA), organised a direct-action strategy; to provide treatment and undertake/coordinate research by a team of 200 scientists and an expenditure of 70 million USD per year, and this led to a confrontation with the then Director-General, Hiroshi Nakajima of Japan (Mann, 1987). Because of this confrontation, Mann left the WHO, and the United States and other countries decided to pull out GPA from the WHO (Krim, 1998; Merson & Inrig, 2018). After some years of discussion and debate, UNAIDS was founded in 1994–1995 under the leadership of Peter Piot (Fee & Parry, 2008).

The Global Fund to Fight AIDS, tuberculosis and malaria (the Global Fund), was created in 2002 as an innovative financing mechanism that seeks to rapidly raise and disburse funding for programmes that reduce the impact of HIV/AIDS, tuberculosis and malaria in low- and middle-income countries (Schocken, 2021). The idea of the

This chapter is largely taken from: Velásquez, G. (2020 November). *The World Health Organization Reforms in the Time of COVID-19*. South Centre Research Paper 121. https://www.southcentre.int/wp-content/uploads/2020/11/RP-121-rev2.pdf.

G. Velásquez, *Vaccines, Medicines and COVID-19*, SpringerBriefs in Public Health, https://doi.org/10.1007/978-3-030-89125-1_6

Global Fund came from the Brundtland administration, which conceived it as an innovative mechanism to fund the WHO. In this context the Brundtland administration called for a 'Massive Attack on Diseases of Poverty' in December 1999 (the WHO Commission on Macroeconomics and Health, 2001). The Global Fund was finally established in January 2002, outside the WHO, following negotiations involving donors, country governments, non-governmental organisations (NGOs), the private sector, and the United Nations (Every CRS report, 2006).

The Expanded Programme on Immunization was launched by the World Health Assembly in 1974. Gavi, an alliance of public and private sector organisations, institutions and governments, the Bill & Melinda Gates Foundation, UNICEF, the World Bank, the WHO, vaccine manufacturers, NGOs, and research and technical health institutes, was established at the Proto-Board Meeting in Seattle on 12 July 1999. Again, an initiative developed within the WHO to support the global immunisation programme was created outside the WHO.

Unitaid, an initiative of the Governments of France and Brazil, was created in 2006 with the support of Chile, Norway and the United Kingdom. This innovative financing initiative is hosted by the WHO but is an independent agency that operates autonomously.

COVAX is the vaccines pillar of the WHO Access to COVID-19 Tools (ACT) Accelerator, formally known as 'the COVID-19 Vaccines Global Access Facility'. It was created in April 2020 and is co-led by Gavi, the Coalition for Epidemic Preparedness Innovations (CEPI), and the WHO. Funding and the power to act are, once again, outside the WHO.

It seems that at every health crisis, whether it is AIDS, vaccines, or COVID-19, the WHO member countries opted to allocate the funding and the power to act outside the WHO.

In the current unprecedented health crisis caused by COVID 19, some industrialised countries seem to have become aware of the structural problems of the WHO, as set out in a 'non-paper' presented in August 2020 by France and Germany (Governments of France and Germany, 2020), or as reflected in the intervention of the President of Switzerland at the World Health Assembly in May 2020 (Swiss Federal Council, 2020). Other suggestions were presented in September 2020 by Chile (together with Uruguay, Paraguay and Bolivia) and the United States. These last two proposals will not be analysed in this chapter as they only refer to the process and methodology for the review of the International Health Regulations (IHR) and of the scope and transparency of the WHO pandemic declarations of a public health emergency of international concern (PHEIC).

This chapter seeks to identify the main problems faced by the WHO in the light of the COVID-19 crisis, and to suggest key elements that a reform of the Organization would need to consider, based on some pertinent proposals of the non-paper presented by France and Germany and in view of the concerns and needs of the countries of the Global South.

6.2   Background

The first major reform of the WHO was led by Halfdan Mahler (Director-General 1973–1988). The Declaration of Alma-Ata, proclaimed at the International Conference on Primary Health Care on 12 September 1978, underlined the urgency of promoting primary health care and access to an acceptable level of health for all (WHO, 1978). Mahler's objective to reach 'Health for All by the Year 2000' significantly changed the orientation of the organisation.

The Director-General of the WHO from 1998 to 2003, Gro Harlem Brundtland, made the most important reform of the organisation after the change of direction brought about by the Alma-Ata conference (1978). A reform described by many as neoliberal, Brundtland initiated what has been termed the 'privatization of the WHO' (Chorev, 2013; Velásquez, 2016). The call 'we must reach out to the private sector' was launched by Brundtland at her first World Health Assembly (Brundtland, 1998).

In May 2011, a few months before the end of her first mandate, Margaret Chan (Director-General, 2007–2017) launched, in her own words, that the 'WHO is now embarking on the most extensive administrative, managerial, and financial reforms, especially financial accountability, in its 63-year history' (World Health Assembly, 64 & Chan, 2011). An ambiguous and disjointed reform that in the 5 years of her second and last mandate did not manage to conclude on the most urgent and controversial issues such as the issue of non-state actors. Tedros Adhanom Ghebreyesus, elected Director-General of the WHO in 2017, announced in his opening speech to the first Executive Board (January 2018) a plan to transform the WHO. The transformation plan was interrupted by the arrival of COVID-19 in December 2019.

On 31 December 2019, Chinese authorities reported several dozen cases of pneumonia from an unknown cause. On 20 January 2020, the WHO reported the first confirmed cases in China, Thailand, Japan and South Korea, and on 30 January 2020, the Director-General declared the novel coronavirus outbreak a public health emergency of international concern (PHEIC), the highest WHO alarm level (WHO, 2020b).

In a context of criticism, mainly from the US Government, of the WHO handling of the pandemic, particularly on the reasons for an alleged delay in announcing the highest level of alarm and the US complaint about China's influence on the announcement of the pandemic, President Trump announced the departure of the United States from the WHO (BBC News Mundo, 2020).

6.3   COVID-19 and the WHO Reform

The COVID-19 pandemic has highlighted the need for a strong and independent global health governing body capable of managing a global health crisis. During the first 6 months of the pandemic there was much talk of what the WHO does or does not do and what it could or could not do. As recently pointed out by Gostin, Moon, and Mason Meier, '[t]he world is facing an unprecedented global health threat, and the response is highlighting structural limitations in the ability of international organisations to coordinate nationalist States' (Gostin et al., 2020).

Faced with the US Government's irresponsible announcement of its withdrawal from the WHO, Germany and France decided to start a process to 'reform the WHO from outside' by presenting, as noted earlier, a document entitled 'Non-Paper on Strengthening WHO's leading and coordinating role in global health, with a specific view on the WHO's work in health emergencies and improving IHR implementation' (hereafter 'the non-paper') (Governments of France and Germany, 2020).

The non-paper is based on the resolution adopted by the 73rd World Health Assembly in May 2020 requesting the Director-General to 'initiate, as soon as possible and in consultation with Member States, a gradual process of impartial, independent and comprehensive evaluation, including by using existing mechanisms, as appropriate, to review the experience gained and lessons learned from the international health response coordinated by the WHO to, inter alia, COVID-19:

(i) the effectiveness of the mechanisms available to the WHO;
(ii) the functioning of the IHR and the status of implementation of relevant recommendations of previous IHR Review Committees;
(iii) the contribution of the WHO to the efforts of the United Nations system as a whole;
(iv) and WHO actions and timetables in relation to the COVID-19 pandemic, and make recommendations to improve global pandemic prevention, preparedness and response capacity, including through strengthening, as appropriate, the WHO Health Emergency Programme' (World Health Assembly, 73, 2020).

Amid the most intense health crisis in the last hundred years, the WHO, as the United Nations specialised agency for health, stands at what probably is the greatest challenge in its history. It is a profound crisis of identity as the Secretariat in Geneva is weakened by the imbalances in international relations reflected in confrontations between some governments of the Global North and the Global South, the United States' withdrawal from the organisation, and the decisive influence of the private and philanthropic sectors in setting its agenda. All this unfortunately leads to an unprecedented loss of credibility in the eyes of the public opinion. This is the challenge facing the WHO today, and countries should see COVID-19 as an opportunity to build a stronger member-led agency, rather than to attack it or allow for a greater influence by the private sector and philanthropy.

In the first half of 2020, the WHO Secretariat was particularly active in providing information, recommendations and guidelines for the management of COVID-19. More than 400 guidance documents for individuals, communities, schools, businesses, industries, health workers, health facilities and governments related to different aspects of the COVID-19 pandemic were produced by the WHO Secretariat in the first 6 months of 2020 (WHO, 2020a). What happened and what is continuing to happen is that some countries did not follow the WHO, however timely and relevant the recommendations were. What is needed today, on the eve of the arrival of a possible vaccine, is a strong, independent organisation capable of supporting countries in tackling problems such as those currently being caused by COVID-19.

Today, more than ever, it is necessary to form a strong coalition of countries willing to defend the public character, authority and independence of the WHO, to allow it to set public health rules at a global level with the capacity and the instruments necessary to put those rules into practice.

According to the non-paper, expectations on the mandate of the WHO are immense. The Organization must set health norms and standards, promote monitoring and implementation in a wide range of health areas, set the research agenda, articulate evidence-based and ethical health policies, react to disease outbreaks around the world, and finally monitor the global health situation.

Unfortunately, to fulfil this mandate, the WHO currently does not have the required legal, financial or structural instruments, says the non-paper. More precisely, it is not that the WHO does not have the instruments to implement its mandate, but rather that it is unable to use them. The high imbalance among Member States assessed financial contributions and the high level of voluntary (public and private) and philanthropic financing, contributes to the problem.

The most logical way to approach a reform process is to start by identifying the problems, so that we know exactly what we want to reform and how we are going to reform it. There are three major problems/issues that a WHO reform would have to address, as explained in the following three points.

6.3.1 Problem 1: The Public-Private Sector Dilemma

The WHO was created in 1948 as a specialised public agency of the United Nations System to improve and maintain health around the world.

For many years, this agency was financed by public funds from regular mandatory contributions by the 194 member countries. Over the past 20 years, voluntary contributions (private or public) have grown rapidly.

The biggest problem of the WHO today, and at the same time the cause of many other ills, as stated in the non-paper, is the loss of control over the regular budget. This has led to a progressive 'privatization' of the agency. 'At the time when WHO's 194 Members States, after lengthy negotiations, adopt the programme budget, it is only partly predictably financed (by roughly 20% of assessed contributions)' (Governments of France and Germany, 2020). Approximately 80% of the budget is in the hands of voluntary (public and private) contributors, including philanthropic entities such as the Bill & Melinda Gates Foundation and a small group of industrialised countries, which make donations for specific purposes chosen often by them in a unilateral manner. Over-reliance on voluntary contributions (private or public) results in an inability to set priorities based on the global public health priorities. Member States try to set priorities, but funds come for specific issues, selected by a small number of donors who have a decisive role in deciding what the organisation can or cannot do. As the German-French non-paper makes clear: '… the funding coming in is largely based on individual donor interests (…). The current way of funding WHO has led to a high risk of donor dependency and vulnerability…' (Governments of France and Germany, 2020).

It is surprising that specialised agencies of the United Nations System could be increasingly dependent on voluntary contributions (private or public) that make it impossible for the Member States to define global priorities. There is an urgent need for the UN General Assembly to define clear criteria and principles for financing the whole system. Why not define, as a mandatory standard, that at least 51% of the budget must come from assessed contributions by governments? And to preserve the multilateral and democratic nature of the agencies, it would also be urgent to define the maximum percentage (10 or 15%, for example) that a single contributor (private or public) can contribute to the organisation. Currently, there do not seem to be any obstacles preventing a single entity from contributing a large part, even more than 50%, of the WHO budget.

In her speech to the World Health Assembly in May 2020, the Swiss President Simonetta Sommaruga explained that the WHO, which currently depends on voluntary contributions for 80% of its budget, requires sustainable funding to be able to fulfil its important role. She added, 'Let us ask ourselves – is it fair to expect so much from the WHO while funding it in such an arbitrary manner?' (Swiss Federal Council, 2020).

The most urgent reform of the Organization which should be addressed by Member States is not the lack of funding, as some industrialised countries suggest, but how and by whom this agency is funded (European Union, 2020). It is a question of how to progressively recover the public and multilateral character of the institution. This is a fundamental condition for effectively putting the WHO at the service of the global public health. An increase in the regular public budget will enable the WHO to devote itself to the priorities set by all the Member States without having to constantly follow the priorities of an agenda set by the donors.

Closely related to the public/private role of the WHO is the debate known as FENSA (Framework of Engagement with Non-State Actors) 'WHO collaboration with non-state actors' that the Margaret Chan reform left unfinished.

After 5 years of complex and slow negotiations on the WHO reform, the 69th World Health Assembly (2016) approved a resolution on the 'WHO Collaboration with Non-State Actors' as part of the reform initiated by the then Director-General Margaret Chan in 2011. The FENSA process was essentially a debate/negotiation on the nature of the Organization and the role that the private sector would play in it. Talking about the 'private sector' in the context of the WHO is complicated because 'non-state actors' working in health include non-profit non-state actors such as NGOs like Médecins Sans Frontières (MSF). However, the WHO also defines non-state actors as private for-profit entities, such as the pharmaceutical companies, as well as philanthropic foundations, and there are questions whether some of the latter are for-profit or not (Astruc, 2019).

The major point of controversy for the adoption of FENSA was the debate on the definition of a clear policy and mechanisms to avoid the conflicts of interest that could arise in the interaction of the WHO with the private sector, a point on which unfortunately no clear conclusion was reached. A consensus was only achieved to totally exclude funds from the arms and tobacco industry, but the door was left wide open for money from the pharmaceutical industry or certain 'less healthy' industries.

In May 2020, the WHO Director-General announced the creation of the WHO Foundation, an independent grant-making entity that will support the budget of the Organization's efforts to address global health challenges (WHO, 2020c). Based in Geneva, legally separate from the WHO, the foundation will accept contributions to the WHO from the general public, individual major donors, and corporate private partners. The WHO Foundation will simplify the processing of philanthropic contributions in support of the WHO and will accept contributions in support of every aspect of the agency's mission (Philanthropy News, 2020). With the creation of the WHO Foundation as an independent and flexible way to finance the WHO, the imbalance between private and public in the WHO risks getting worse.

6.3.2 Problem 2: The Dilemma Between Voluntary Recommendations and Binding Instruments in the Health Field

A fundamental and historical responsibility of the WHO has been the management of the global action against the international spread of diseases. Under Articles 21(a) and 22 of the WHO Constitution (2006),[1] the World Health Assembly is empowered to adopt regulations 'for the prevention of the international spread of disease', which, once adopted by the Health Assembly, become effective for all the WHO Member States, 'except those which expressly reject them within the time limit'.[2]

The International Health Regulations (IHR) were adopted by the WHA in 1969 and revised in 2005 due to the limitation of the number of mandatory reporting diseases (yellow fever, plague and cholera). The 2005 IHR, while not limiting the number of diseases, placed a limitation on measures that may affect international traffic or trade. The purpose of the IHR (2005) is 'to prevent, protect against, control and provide a public health response to the international spread of disease in ways

---

[1] WHO Constitution Article 21: 'The Health Assembly shall have authority to adopt regulations concerning: (a) sanitary and quarantine requirements and other procedures designed to prevent the international spread of disease; (b) nomenclatures of diseases, causes of death and public health practices; (c) uniform standards of diagnostic procedures for international use; (d) uniform standards of safety, purity and potency of biological, pharmaceutical and similar products in international trade; (e) advertising and labelling of biological, pharmaceutical and similar products in international trade'. WHO Constitution Article 22: 'These regulations shall come into force for all Members after due notice of their adoption by the Health Assembly, except for those Members which shall inform the Director-General of their rejection or reservation within the period specified in the notice'.

[2] Ibid.

that are commensurate with and restricted to public health risks, and which avoid unnecessary interference with international traffic and trade' (WHO, 2005).

In this context, it could be said that in the first half of 2020 many countries acted in violation of the IHR (Bussard, 2020), and the fact that the non-paper and a large part of the interventions of the countries in the Executive Board Special session on the COVID-19 response on 5 October 2020 called for urgent revision of the 2005 IHR (Alas & Ido, 2020), serves as recognition that the tools currently available to the WHO are insufficient.

Paradoxically, while the international trade rules of the World Trade Organization (WTO) are binding, the WHO does not have the legal means to enforce disciplines that are vital for the protection of global health.

6.3.3   Article 19 of the WHO Constitution

Article 19 of the WHO Constitution states: 'The Health Assembly shall have authority to adopt conventions or agreements in respect of any matter within the competence of the Organization. A two-thirds vote of the Health Assembly shall be required for the adoption of such conventions or agreements, which shall come into force for each Member when accepted by it in accordance with its constitutional processes' (WHO, 2006).

In May 2012, the World Health Assembly adopted a resolution that sought to change the dominant WHO model of 'recommending' (Correa, 2016). This resolution aimed to introduce an alternative model to the Research and Development (R&D) model for pharmaceuticals by calling for the initiation of negotiations for a binding international treaty as a means of funding research for medicines.

A binding global treaty or convention, negotiated in the WHO, could enable the sustainable financing of research and development of useful and safe drugs at prices affordable to the population and public social security systems. The adoption of such a convention within the framework of the WHO, based on Article 19 of its constitution, could also make it possible to review the way in which the WHO operates in a broader sense. The negotiation of 'global and binding instruments on health matters of global concern' is perhaps the most promising avenue for the role that the WHO could take on in the future (Correa, 2016).

In its entire history, the WHO has only once used Article 19 of its constitution to negotiate a convention of a binding nature. In May 2003, after 3 years of negotiations and 6 years of work, the World Health Assembly unanimously adopted the WHO Framework Convention on Tobacco Control, which has now been signed by 177 countries. This was the first – and so far, the only – time that the WHO exercised the power to adopt an international treaty in a substantive area to provide a legal response to a global health threat.

The Framework Convention on Tobacco Control (FCTC) enabled the 177 signatory countries to progressively approximate their legislation to address the problem of smoking. The treaty does not set out agreed standards but also encourages the parties to adopt stricter measures through laws and regulations passed by the parliaments or other competent national bodies. This is undoubtedly one of the greatest achievements of the WHO in its entire history. Why not build on this successful example?

The recommendation to launch negotiations on an agreement on R&D for medicines has not been able to move forward because of lack of a wide support among the WHO members and the opposition from the industrialised countries where the powerful pharmaceutical industry is located. The crisis caused by COVID-19 is a historic opportunity to revisit this issue and help to recover the credibility of the Organization.

6.3.4 Problem 3: The Dilemma Between Regulations and Humanitarian Aid

Another important issue that must be addressed is the dilemma between a standard-setting body responsible for the formulation and creation of standards and instruments, including of a binding nature, and for the administration of international health regulations, versus an agency responsible for providing humanitarian assistance in cases of health emergencies, thereby competing with and duplicating the efforts of other agencies such as the Global Fund, Gavi (including the COVAX facility), Unitaid, other UN agencies such as UNICEF, UNAIDS or UNDP, and large NGOs such as MSF.

In fact, the WHO handling of global health emergencies has not been the most brilliant in recent years. Was H1N1 an industry operation, a false pandemic as Director-General Margaret Chan herself asked, reflecting the criticisms that many observers and countries made at the time: 'First, did the WHO make the right call? Was this a real pandemic or not? And second, were WHO decisions, advice, and actions shaped in any way by ties with the pharmaceutical industry? In other words, did the WHO declare a fake pandemic to line the pockets of industry?' (World Health Assembly, 64 & Chan, 2011).

As a result of the mistakes made by the WHO in managing the H1N1 influenza epidemic, Zika and Ebola, there has been a trend in recent years to strengthen the role of the WHO in emergency and humanitarian work. The French-German non-paper also suggests strengthening work in emergencies. This would give the Organization a dual mission: a normative one and a humanitarian one. However, there are many who believe that the WHO should prioritise its normative functions and leave humanitarian health work to other agencies (Gostin et al., 2015; Yach, 2016; Clift, 2013).

The coordination by the WHO of actors such as Gavi (including the COVAX facility), CEPI, and the Global Fund, with significantly larger budgets and managed with the participation of the private sector, is illusory, as the difficulties in organising the arrival of future vaccines for COVID-19 are showing.

Member countries of the WHO and its Secretariat will have to choose between a management office for projects primarily financed by the private and philanthropic sectors, or the reconstruction of an independent public international agency to promote, preserve and regulate health by recommending or setting norms, strategies and standards. This is a key dilemma for the WHO.

A choice will have to be made between what a few donors want the WHO to be or do, and what the world needs today from a United Nations agency dedicated to health. For those who still believe that the United Nations must play a leading role in the area of health, and even more so for those who want to offer solutions and contribute to the reform of the WHO, the COVID-19 pandemic will perhaps be the last chance for this agency.

6.4   The International Health Regulations (IHR)

The IHR (2005) is an international agreement signed by 196 countries,[3] including all Member States of the WHO. Its aim is to help the international community prevent and respond to serious public health risks that may cross borders and threaten the world's population. The purpose and scope of the IHR (2005), which entered into force in 2007, is 'to prevent, protect against, control and provide a public health response to the international spread of disease in ways that are commensurate with and restricted to public health risks, and which avoid unnecessary interference with international traffic and trade' (WHO, 2014).

The purpose of the revised IHR (2005) is to 'prevent, protect against, control and provide' a response to any public health emergency of international concern (PHEIC) (Article 2 IHR).

Ebola in 2014 and Zika in 2016 were both regarded as PHEICs: they were considered extraordinary events which created public health risks for other states and required a coordinated international response (Article 1 IHR). COVID-19 is the most recent and severe case of PHEIC ever dealt with by the WHO. During a PHEIC, the WHO Director-General may issue temporary recommendations. However, due to their character as 'non-binding advice' (Article 1 IHR), States may follow them or not. The temporary recommendations issued during the Ebola crisis, for instance, were widely ignored with devastating effects.

---

[3]The 194 Member States plus two non-member states of the WHO – The Holy See and Liechtenstein.

6.4.1 Taking a Straightforward Approach: Modifying the IHR

The easiest way to address one of the problems that must be addressed by the WHO reform is, obviously, to modify the IHR. Only one single word needs to be cut: 'Art. 1 IHR could be modified to the extent that temporary recommendations are defined as "binding" measures. In light of recent state practice this approach seems, however, to be out of question' (Frau, 2016).

The non-paper rightly points out that 'While other global legally binding instruments include incentive mechanisms for implementation and reporting, the IHR does not currently provide for such mechanisms' (Governments of France and Germany). This means that the capacity of the WHO Secretariat is quite limited and depends on the goodwill of countries to cooperate. Other binding legal frameworks, such as the WTO trade agreements, include specific notification and transparency procedures that allow its members to monitor the extent to which other members comply with their obligations. In addition, the WTO rules provide that a member that fails to conform its conduct to any of the obligations of the agreements covered by the organisation may suffer suspension of trade benefits. In common parlance, this consequence is called 'trade sanctions' (CEPAL, 2013).

Article 21 of the WHO Constitution states that the Health Assembly has the authority to adopt regulations concerning, inter alia, sanitary and quarantine requirements and other procedures designed to prevent the international spread of disease. Article 22 stipulates that 'Regulations adopted pursuant to Article 21 shall come into force for all Members after due notice has been given of their adoption by the Health Assembly except for such Members as may notify the Director-General of rejection or reservations within the period stated in the notice' (WHO, 2006). However, there are no mechanisms to enforce the adopted regulations if not complied with by members that have not rejected them or made reservations. This is the gap that must be addressed to empower the WHO to effectively protect the global public health in case of a PHEIC.

6.5 Non-paper Proposals of Action

The reform proposed in the non-paper contains 10 actions, of which several are highly relevant (Governments of France and Germany, 2020).

Action 1: Consider a general increase of assessed contributions. This proposal by France and Germany is a major step in the debate on the WHO reform. Admitting that the Organization must be a public entity is the first condition for any coherent reform of the WHO. For more than 20 years, the regular budget of the WHO has been frozen by the United States and other industrialised countries that demanded zero growth.

In the early 1980s, the WHA introduced a 'zero-real growth policy' for the regular budget. This policy froze membership dues in real dollar terms so that only inflation and exchange rates would influence members' assessed contributions. In 1993, the WHA voted for a more stringent budgetary policy, moving the Organization from 'zero real growth' to 'zero nominal growth' for assessed contributions. This policy shift made the Organization increasingly reliant on extra budgetary funds (Reddy et al., 2018).

Action 2: Strengthen the normative role of the WHO. In the face of the multiplication of international actors in the health field, strengthening the normative capacity of the WHO is a way to give it back its identity and specificity and to allow other public-private actors, philanthropists, to continue to act, while respecting and applying the WHO standards. This second action proposed in the non-paper does not go far enough, as it does not mention what the instruments will be to ensure compliance with the standards that should logically be via Article 19 and developing rules under Articles 21 and 22 of the WHO Constitution.

Action 3: Establish strong and sustainable governance structures that enable WHO Member States to provide adequate oversight and guidance to the work of the WHO in health emergencies. A clear lesson from COVID-19 is that the WHO must have 'strong and sustainable governance structures', but this action is insufficient if strong governance structures are not identified. The non-paper merely mentions that a subcommittee of the Executive Board should be established to monitor health emergencies and crises.

Declarations of the highest level of health crisis (PHEIC) should be accompanied by effective compliance mechanisms to be activated in times of global health crises, e.g. to ensure that pandemic-related diagnostics, treatments and vaccines are accessible and affordable to all.

Actions 8 and 9 of the non-paper refer to the reform of the PHEIC and the transparent implementation of the health regulations at the national level. As already mentioned, the declaration of a PHEIC should be explicitly accompanied by the possibility of using compliance mechanisms based on binding rules. On the transparency of the application of the IHR at the national level, the non-paper calls for improved collaboration and strengthening of the system for reporting outbreaks or PHEICs. The immediate reporting of such problems should be mandatory.

6.6   The Special Meeting of the Executive Board on 5–6 October 2020

In the context of the health crisis caused by COVID-19, the resolution WHA73 'COVID-19 Response' (May 2020) and the non-paper presented by France and Germany, the extraordinary meeting of the Executive Board on 5–6 October 2020 became a kind of forum on how to address the reform of the WHO.

At the extraordinary meeting, several countries referred in their interventions to the non-paper presented a few weeks earlier; it, hence, became an important element of a diplomatic strategy aiming at starting a debate on the WHO reform.

The second day of the extraordinary meeting was dedicated to the review of the progress of two committees and one panel that are in charge of the implementation of resolution WHA73: COVID-19 Response, of May 2020 (World Health Assembly, 73, 2020).

1. The Independent Panel for Pandemic Preparedness and Response (IPPR).
2. The IHR Review Committee
3. The Independent Oversight and Advisory Committee for the WHO Health Emergencies Programme (IOAC).

The two committees and the panel are composed of recognised international experts appointed by the WHO Director-General. Proposals for the implementation of resolution WHA73, now assumed to be part of a WHO reform, are expected to go to these bodies, or at least to the second 'IHR Review Committee' that is considering the IHR review.

6.7   Concluding Remarks

The three main problems identified in relation to the inability of the WHO to respond to situations such as the one posed by the COVID-19 crisis require a discussion of:

1. The public nature and role of the WHO.
2. The absence of binding mechanisms for the enforcement of its directives, norms and standards.
3. The dilemma between the normative and the humanitarian role of the WHO.

A reform of the WHO that aims to respond to the existing structural problems should then define mechanisms to progressively regain the public character of the Organization, so as to control at least 51% of the budget in a period of, for instance, 7 years. This means that the regular mandatory assessed contributions of the Member States should represent at least 51% of the agency's total budget.

Effective coordination by the WHO of global health issues requires the use of Articles 19, 20 and 21 of its Constitution for the approval of binding instruments and compliance mechanisms that ensure the effective application of directives, regulations and standards issued by the Organization.

The third point of the reform is perhaps the most complex and controversial – the dilemma between the normative and the humanitarian role of the WHO. For the reasons explained throughout this book, and considering the multiplication of actors addressing health issues and the mistakes and delays in the management of previous epidemics (H1N1, Zika, Ebola), the WHO should, as a priority, concentrate on its normative work.

The more than 400 high-quality and relevant documents produced by the WHO during the first 6 months of 2020 are a clear and positive sign of what this agency can do. If the tools and instruments were found to make the relevant standards enforceable, the world would be much better off.

References

Alas, M., & Ido, V. (2020). WHO's Executive Board assesses current COVID-19 response and requests to be more involved in the review processes. SOUTHNEWS, No. 346, South Centre, 20 October 2020. https://us5.campaign-archive.com/?u=fa9cf38799136b5660f367ba6&id=a2e651d8f1.

Astruc, L. (2019). *L'art de la fausse générosité: La fondation Bill et Melinda Gates*. Actes Sud Editions.

BBC News Mundo. (2020). Estados Unidos se retira de la OMS: Trump notifica oficialmente a Naciones Unidas de la salida de su país. 7 July 2020. https://www.bbc.com/mundo/noticias-internacional-53329647.

Brundtland, G. H. (1998). Speech to the Fifty-first World Health Assembly Geneva, 13 May 1998, A51/DIV/6 13. https://apps.who.int/gb/archive/pdf_files/WHA51/eadiv6.pdf.

Bussard, S. (2020). La plupart des Etats ont violé le texte fondamental de l'OMS. *Le Temps*, 15 June 2020.

CEPAL. (2013). Tipologia de instrumentos internacionales. LC/L.3719, 2013 https://www.cepal.org/es.

Chan, M. (2011). Opening address to the Executive Board by Dr Margaret Chan, Director-General, special session on WHO reform. Geneva, Switzerland, 1 November 2011. https://apps.who.int/gb/ebwha/pdf_files/EBSS/EBSS2_ID9-en.pdf.

Chorev, N. (2013, August). Restructuring neoliberalism at the World Health Organization. *Review of International Political Economy, 20*(4). https://doi.org/10.1080/09692290.2012.690774

Clift, C. (2013). The Role of the World Health Organization in the International System, Centre on Global Health Security Working Group Papers, Working Group on Governance | Paper 1. London: Chatham House. https://www.chathamhouse.org/sites/default/files/publications/research/2013-02-01-role-world-health-organization-international-system-clift.pdf.

Correa, C. M. (2016). Una Resolución de la Asamblea Mundial de la Salud. Curar la enfermedades de los pobres? *Le Monde Diplomatique*, February 2016. https://mondiplo.com/curar-por-fin-las-enfermedades-de-los-pobres.

European Union. (2020). European Union Statement at WHO Special Session of the Executive Board on the COVID-19 response. 5–6 October 2020. https://apps.who.int/gb/statements/EBSS5/PDF/EU_4.pdf.

Every CRS Report. (2006). The global fund to fight AIDS, Tuberculosis, and Malaria: Background 2003–2006. https://www.everycrsreport.com/reports/RL31712.html.

Fee, E., & Parry, M. (2008). Jonathan Mann, HIV/AIDS, and human rights. *Journal of Public Health Policy, 29*(2008), 54–71. https://link.springer.com/article/10.1057/palgrave.jphp.3200160#citeas

Frau, R. (2016). Creating legal effects for the WHO's International Health Regulations (2005) – Which way forward? *Völkerrechtsblog*, 13 April 2016. DOI: https://doi.org/10.1717 6/20171201-133647.

Gostin L., Moon, S., & Mason Meier, B. (2020, October). Reimagining global health governance in the age of COVID-19. *American Journal of Public Health, 110*(11), 615–619. https://doi.org/10.2105/AJPH.2020.305933.

Gostin, L., Sridhar, D., & Hougendobler, D. (2015). The normative authority of the World Health Organization. *Public Health*, forthcoming. SSRN: https://ssrn.com/abstract=2634181.

Governments of France and Germany. (2020). Non-paper on strengthening WHO's leading and coordinating role in global health. With a specific view on WHO's work in health emergencies and improving IHR implementation, August 2020. http://g2h2.org/wp-content/uploads/2020/08/Non-paper-1.pdf.

Krim, M. (1998, October). Jonathan Mann 1947–1998. *Nature Medicine, 4*, 1101. https://doi.org/10.1038/2592.

Mann, J. M. (1987, December). The World Health Organization's global strategy for the prevention and control of AIDS. *The Western Journal of Medicine, 147*(6), 732–734.

Merson, M., & Inrig, S. (2018). End of the global programme on AIDS and the launch of UNAIDS. In *The AIDS pandemic: searching for a global response*. Springer. https://doi.org/10.1007/978-3-319-47133-4_16

Philanthropy News. (2020). WHO establishes foundation to help address global health challenges. *PND*, 31 May 2020. https://philanthropynewsdigest.org/news/who-establishes-foundation-to-help-address-global-health-challenges.

Reddy, S., Mazhar, S., & Lencucha, R. (2018). The financial sustainability of the World Health Organization and the political economy of global health governance: A review of funding proposals. *Globalization and Health, 14*, 119. https://doi.org/10.1186/s12992-018-0436-8

Schocken, C. Overview of the global fund to fight AIDS, Tuberculosis and Malaria. https://www.cgdev.org/sites/default/files/archive/doc/HIVAIDSMonitor/OverviewGlobalFund.pdf. Accessed 26 June 2021.

Swiss Federal Council. (2020). Press release, President Sommaruga takes part in 73rd World Health Assembly. 18 May 2020. https://www.admin.ch/gov/en/start/documentation/media-releases.msg-id-79150.html.

Velásquez, G. (2016). Qué remedios para la Organizacion Mundial de la Salud. *Le Monde Diplomatique* No. 253, November 2016.

WHO. (1978). Declaration of Alma-Ata. International conference on Primary Health Care, Alma-Ata, USSR, 6–12 September 1978. https://www.who.int/publications/almaata_declaration_en.pdf.

WHO. (2005). International Health Regulations (2005). https://www.who.int/publications/i/item/9789241580496

WHO. (2006). Constitution of the World Health Organization, Articles 19 and 21. https://www.who.int/governance/eb/who_constitution_en.pdf.

WHO. (2014). Emergencies: Ten things you need to do to implement the International Health Regulations – Know the IHR; Purpose, scope, principles and concepts. World Health Organization, 26 May 2014. https://www.who.int/news-room/q-a-detail/emergencies-ten-things-you-need-to-do-to-implement-the-international-health-regulations.

WHO. (2020a). Director-General's opening remarks at Executive Board Meeting. Special Session on the COVID-19 Response, 5 October 2020. https://www.who.int/dg/speeches/detail/who-director-general-s-opening-remarks-at-executive-board-meeting.

WHO. (2020b). Timeline of WHO's response to COVID-19. https://www.who.int/news-room/detail/29-06-2020-covidtimeline.

WHO. (2020c). WHO foundation established to support critical global health needs. Geneva, 27 May 2020. https://www.who.int/news-room/detail/27-05-2020-who-foundation-established-to-support-critical-global-health-needs.

WHO Commission on Macroeconomics and Health. (2001). Macroeconomics and health: Investing in health for economic development. Report of the Commission on Macroeconomics and Health. World Health Organization. https://apps.who.int/iris/handle/10665/42435.

World Health Assembly, 64, & Chan, M. (2011). *Address by Dr Margaret Chan, Director-General to the Sixty-fourth World Health Assembly*. World Health Organization. https://apps.who.int/iris/handle/10665/2124

World Health Assembly, 73. (2020). WHA73 draft resolution on COVID-19 response, May 2020. https://apps.who.int/gb/ebwha/pdf_files/WHA73/A73_CONF1Rev1-en.pdf.

Yach, D. (2016, November). World Health Organization reform—A normative or an operational organization? *American Journal of Public Health*, *106*(11): 1904–1906. https://ajph.aphapublications.org/doi/full/10.2105/AJPH.2016.303376.

Epilogue

The COVID-19 crisis has demonstrated the urgent need for a multilateral health body such as the World Health Organization (WHO). The epidemic has highlighted the fragility of the global system, the inequalities in the face of the health crisis and the need to strengthen the multilateral system. The multilateral system is vital for defining and coordinating responses to health challenges. This is especially true in the context of the current pandemic and others that may emerge.

How can we ensure that this multilateral body can exercise its authority and take the lead? How can the countries of the Global South contribute to strengthening the agency so that it can better serve the interests of their populations, who are particularly affected by inequalities in access to health, medicines and vaccines in the aftermath of the COVID-19 crisis?

This book, which brings together the reflections produced by the South Centre between 2020 and early 2021, is a contribution to the debate and offers an analysis for policymakers, researchers and different stakeholders. We summarise here our conclusions on the following issues, which we have addressed in this book, particularly relevant to access to medicines and vaccines in the context of COVID-19.

On research and development (R&D) of vaccines and medicines and intellectual property

We believe that the management of a pandemic cannot be left to commercial companies competing with the primary intention of making money. The public interest needs to be placed far above commercial interests and knowledge needs to be in the public domain for the advancement of science. It is undeniable that in the last decade there has been progress in recognising the importance of the impact of intellectual property on access to medicines. It is now part of the debate on access to health and also part of the debate on universal health coverage (UHC). We argue that the WHO is the most appropriate multilateral organisation to launch an R&D strategy for universal access to medicines to achieve UHC. While the need to create a binding global instrument for health R&D and innovation has been formulated for

© SC: South Centre 2022

G. Velásquez, *Vaccines, Medicines and COVID-19*, SpringerBriefs in Public Health, https://doi.org/10.1007/978-3-030-89125-1

some years, the need to make it a reality is becoming increasingly urgent in the context of COVID-19. Article 19 of the WHO Constitution would allow WHO Member States to initiate negotiations to create such a binding global instrument for R&D. To be effective, it would have to be capable of being effective, it would have to be capable of being used as an instrument for the development of a global instrument for health R&D and innovation. To be effective, it would need to be able to prioritise R&D according to health needs, coordinate R&D to avoid unnecessary duplication, and devise sustainable public R&D funding mechanisms. A binding international convention on R&D should be negotiated among member countries and implemented, so that intellectual property does not hinder access to medicines and vaccines for all. With this instrument, the world would be better prepared for a health crisis such as the one created by the COVID-19 pandemic.

On global and local manufacturing of medical products

We have shown how, with the COVID-19 pandemic, developed countries have renewed their efforts to increase autonomy in the manufacture of pharmaceuticals and how this has given rise to nationalist approaches. At the same time, it has become clear that, although several COVID-19 vaccines have been successfully developed, there is insufficient global manufacturing capacity to produce the billions of doses needed to protect the world's population. In this context, we argue that there is an urgent need to reopen the debate on local pharmaceutical production and on how developing countries can increase their capacity to participate in the global market for active pharmaceutical ingredients (APIs) and pharmaceuticals, including biologics. The context of increasing pressure on demand for COVID-19 vaccines has brought this issue back to the forefront, while pressure on pharmaceuticals in the Global South remains high and new epidemics may occur.

On access to vaccines and intellectual property

The COVID-19 crisis has acutely illustrated the inequality in access to vaccines, especially because of patents on products that should be considered global public goods. In global emergencies such as that caused by COVID-19, flexibilities such as those provided for in the World Trade Organization's (WTO) TRIPS agreement should be used. For more than 15 years, the WHO and academic research have shown that pharmaceutical patents can be, and often have been, an obstacle to access to vaccines and medicines, and that there are solutions to this dramatic problem.

On patents and access to vaccines and medicines

We believe that the flexibilities allowed by the WTO Trade-Related Aspects of Intellectual Property Rights (TRIPS) agreement should be used, allowing for better production and purchase of vaccines and medicines in case of a health crisis. Patents are a tool to promote innovation, but they should never go against the public interest. The goal of saving lives must always come before commercial interests, however legitimate they may be.

On the necessary reforms of the World Health Organization

We believe that the WHO's inability to respond to situations such as the COVID-19 crisis is linked to three major problems, which need to be discussed by all stakeholders: (1) a certain breakdown in the public character of the WHO; (2) the absence of binding mechanisms for the implementation of its guidelines, norms and standards; and (3) the tension between the normative and humanitarian functions of the WHO. Among other necessary elements of WHO reform, we propose the following reflections that would help to address these three major problems identified.

To progressively restore the public character of the organisation, mechanisms should be defined and put in place to control at least 51% of the budget, over a period of, say, 7 years. This means that regular compulsory contributions from Member States should represent at least 51% of the agency's total budget.

To ensure the effective implementation of directives, regulations and standards issued by the organisation, the WHO should activate the use of Articles 19, 20 and 21 of its constitution that would allow it to put in place binding instruments and mechanisms.

Resolving the tension, or dilemma, between the normative and humanitarian functions of the WHO is perhaps the most complex and controversial of the important aspects of the reforms we consider necessary. For the reasons discussed throughout Chap. 5, and given the multiplication of actors dealing with health issues as well as the mistakes and delays observed in the management of past epidemics (H1N1, Zika, Ebola), we believe that the WHO should focus primarily on its normative work.

We strongly advocate that in today's world of multiple crises – health, social, economic and political – a multilateral health body such as the WHO is essential. Only a strong and democratic multilateral body can defend the interests of the populations of all countries, not just the most powerful and wealthy. Although the WHO is not without its critics – as we have pointed out in this book – the fact remains that it exists, that it has more than 70 years of experience, and that it has extensive global experience and expertise. It is in the interests of the countries of the Global South to bring about the necessary transformations to strengthen WHO action and authority to defend global public goods and interests, first and foremost health for all. Voltaire said that 'if God did not exist, we would have to invent him'. To paraphrase the philosopher, we could say that if the WHO did not exist, it would have to be invented.

Index